Introductory Circuit Analysis

Introductory Circuit Analysis

Hudson Elliott

NY RESEARCH PRESS
New York

Published by NY Research Press
118-35 Queens Blvd., Suite 400,
Forest Hills, NY 11375, USA
www.nyresearchpress.com

Introductory Circuit Analysis
Hudson Elliott

International Standard Book Number: 978-1-64725-424-7 (Hardback)

Cataloging-in-Publication Data

Introductory circuit analysis / Hudson Elliott.
p. cm.
Includes bibliographical references and index.
ISBN 978-1-64725-424-7
1. Electric circuit analysis. 2. Electric network analysis. I. Elliott, Hudson.
TK454 .E54 2023
621.319 2--dc23

Contents

Preface

Every book is initially just a concept; it takes months of research and hard work to give it the final shape in which the readers receive it. In its early stages, this book also went through rigorous reviewing. The notable contributions made by experts from across the globe were first molded into patterned chapters and then arranged in a sensibly sequential manner to bring out the best results.

Circuit analysis is the process of finding all the currents and voltages in each element of an electrical or electronic circuit. Informally, circuit analysis is also known as solving a circuit. The different components of a circuit are resistors, transistors, capacitors, inductors and diodes. Circuit analysis deals with the calculation of unknown electrical circuit quantities such as voltage, current, resistance, impedance, power, etc. There are two important circuit analysis laws also known as Kirchhoff's laws. These laws are the Kirchhoff's Current Law (KCL) and the Kirchhoff's Voltage Law (KVL). KCL is one of the fundamental laws used for circuit analysis which states that the algebraic sum of all currents entering and exiting a node must be equal to zero. KVL states that the directed sum of the potential differences (voltages) around any closed loop is zero. This book provides a comprehensive understanding to the fundamental concepts of circuit analysis. Coherent flow of topics, student-friendly language, and extensive use of examples make it an invaluable source of knowledge for all the readers.

It has been my immense pleasure to be a part of this project and to contribute my years of learning in such a meaningful form. I would like to take this opportunity to thank all the people who have been associated with the completion of this book at any step.

Hudson Elliott

1

Basic Circuits Analysis and Network Topology

1.1 Ohm's Law - Kirchhoff's Law

Ohm's Law

It states that the current through a conductor between two points is directly proportional to the potential difference or voltage across the two points and inversely proportional to the resistance between them.

Mathematical equation that describes this relationship is given as:

$$I = V / R$$

Where:

I - Current through the resistance in units of amperes.

V - Potential difference measured across the resistance in units of volts.

R - Resistance of the conductor in units of ohms.

Ohm's law states that R in this relation is constant and it is independent of the current.

Limitations of ohm's law:

- It is not applicable to non-linear devices such as diodes, Zener diode, voltage regulators etc.
- It is not suitable for non-metallic conductors such as silicon carbide.

Kirchhoff's Law

Kirchhoff's laws are the basic analytical tools to obtain the solutions of voltages and currents for any electric circuit whether it is supplied from a direct current system or an alternating current system.

Kirchhoff's First Law or Kirchhoff's Current Law (KCL)

Total charge or current entering a junction or node is exactly equal to the charge leaving the node as it has no other place to go except to leave as there is no charge lost within the node.

KCL states that, at any node (junction) in a circuit, the algebric sum of currents entering and leaving a node at any instant of time is equal to zero.

$$I_{(Exiting)} + I_{(Entering)} = 0$$

This idea by Kirchhoff is known as the conservation of charge. At any instant of time, the algebric sum of current at a junction or node is zero. The algebric sum of current entering into a junction is equal to the current leaving the junction.

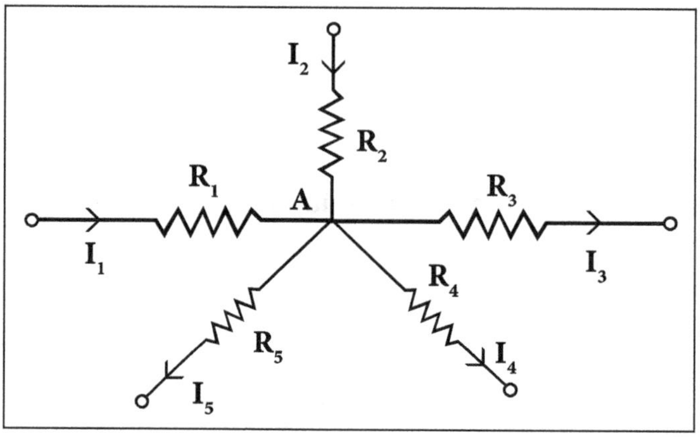

Kirchhoff's current law.

$$I_1 + I_2 = I_3 + I_4 + I_5$$

$$I_1 + I_2 - I_3 - I_4 - I_5 = 0$$

Kirchhoff's Second Law or Kirchhoff's Voltage Law (KVL)

In any closed loop network, the total voltage around the loop is equal to the sum of all the voltage drops within the same loop which is also equal to zero.

KVL states that, in a closed circuit, the algebric sum of all the source voltages must be equal to the algebric sum of all the voltage drops. This idea by Kirchhoff is known as the conservation of energy.

The algebric sum of potential around a closed path is equal to zero. In other words, the sum of potential rise is equal to the sum of potential drop in a closed circuit.

$$V = V_1 + V_2 + V_3$$

$$V - V_1 - V_2 - V_3 = 0$$

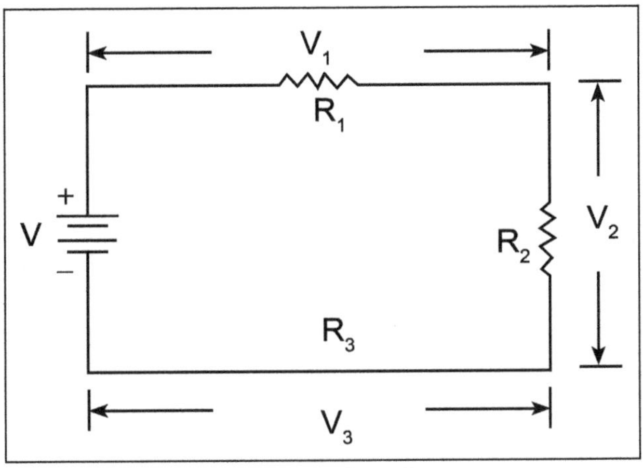

Kirchhoff's voltage law.

Steps to solve a problem using KCL/KVL method:

- Step 1: The branch currents of each branch using KCL was found.
- Step 2: Equations for each loop in terms of unknown currents is obtained by using KVL.
- Step 3: Simultaneous equations obtained in step 2 is solved using calculator to get the unknown currents.
- Step 4: Required branch current and power was found.

Problems

1. Let us consider a current of 0.55A is flowing through the resistance of 10 Ω. Let us determine the potential difference between its ends.

Solution:

Given:

Current, I = 0.55A

Resistance, R = 10 Ω

Formula to be used:

$$V = IR$$

$$= 0.55 \times 10$$

$$= 5.5V$$

2. A supply voltage of 200 V is applied to a 100 Ω resistor. Let us determine the current flowing through it.

Solution:

Given:

Voltage, V = 200V

Resistance, R = 100Ω

Formula to be used:

Current, $I = V/R$

$= 200/100$

$= 2\text{ A}$

3. Let us calculate the resistance of the conductor if a current of 2 A flows through it when the potential difference across its ends is 12 V.

Solution:

Given:

Current, I = 2A

Potential difference, V = 12V

Formula to be used:

Resistance, $R = V/I$

$= 12/2$

$= 6\text{ohm}$

4. Let us determine the current flowing in the 40Ω resistor R_3.

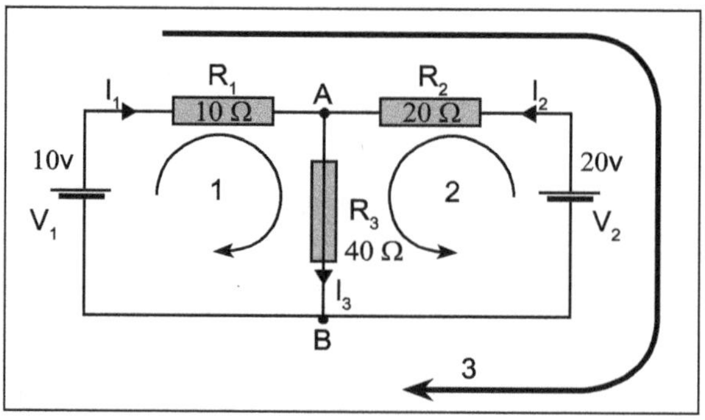

Solution:

Given:

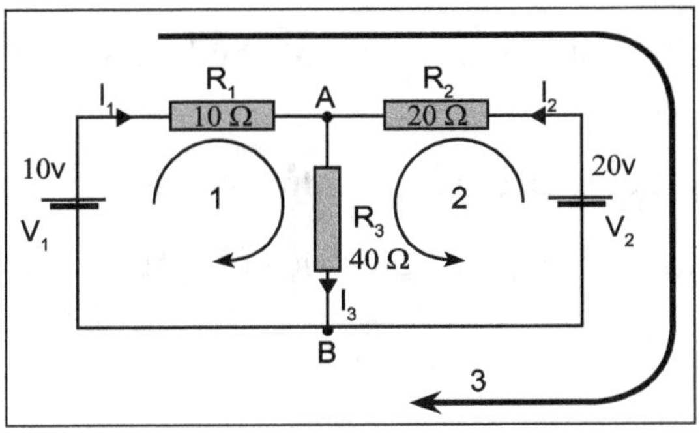

The circuit has three branches with two nodes (A and B) and two independent loops.

Formula to be used:

Using Kirchhoff's current law (KCL), the equations are given as,

Node A:

$$I_1 + I_2 = I_3$$

Node B:

$$I_3 = I_1 + I_2$$

Using Kirchhoff's voltage law (KVL), the equations are given as:

Loop 1 is given as:

$$10 = R_1 \times I_1 + R_3 \times I_3 = 10I_1 + 40I_3 \qquad ...(1)$$

Loop 2 is given as:

$$20 = R_2 \times I_2 + R_3 \times I_3 = 20I_2 + 40I_3 \qquad ...(2)$$

Loop 3 is given as:

$$10 - 20 = 10I_1 - 20I_2 \qquad ...(3)$$

I_3 is the sum of $I_1 + I_2$

We can rewrite the equations as:

Equation (1):

$$10 = 10I_1 + 40(I_1 + I_2) = 50I_1 + 40I_2$$

Equation (2):

$$20 = 20I_1 + 40(I_1 + I_2) = 40I_1 + 60I_2$$

We now have two simultaneous equations that can be reduced to give us the value of both I_1 and I_2.

Substitution of I_1 in terms of I_2 gives us the value of I_1 as -0.143 A.

Substitution of I_2 in terms of I_1 gives us the value of I_2 as +0.429 A.

$$I_3 = I_1 + I_2$$

Current flowing in resistor R_3 is given as:

$$-0.143 + 0.429 = 0.286A$$

Voltage across the resistor R_3 is given as:

$$0.286 \times 40 = 11.44V$$

5. Let us determine the current through a 20Ω resistance and current through a 40Ω resistance.

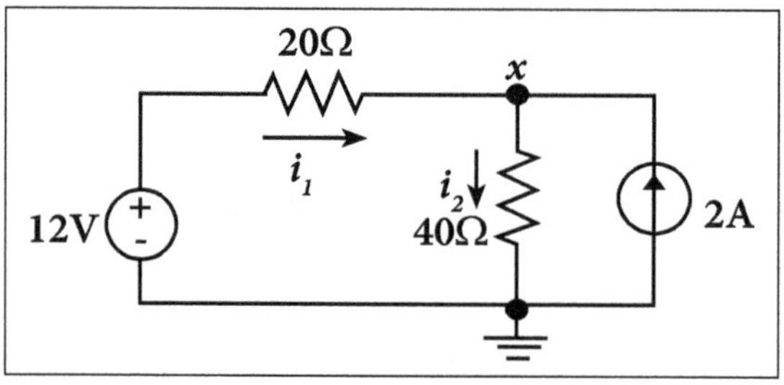

Solution:

Given:

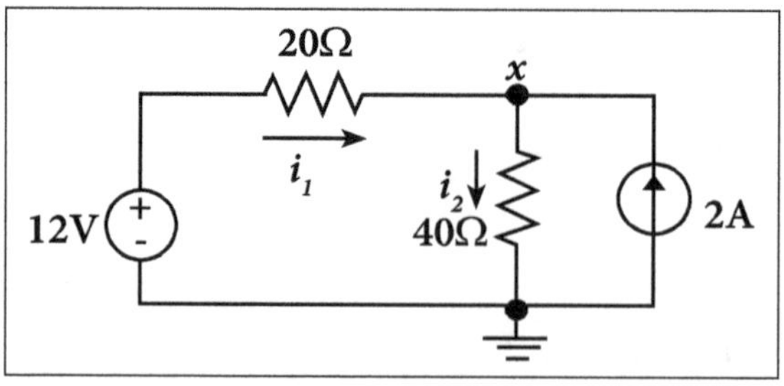

By writing KCL at node x, we have:

$$-i_1 + i_2 - 2A = 0$$

Writting v_x in the circuit using Ohm's law, we have:

$$i_1 = \frac{12\,V - v_x}{20\Omega}, i_2 = \frac{v_x}{40\Omega}$$

$$-i_1 + i_2 - 2A = -\frac{12\,V - v_x}{20\Omega} + \frac{v_x}{40\Omega} - 2A = 0$$

$$-0.6 + 0.05v_x + 0.025v_x - 2 = 0$$

$$v_x = 34.67V$$

Current through a 20Ω resistance is given as:

$$i_1 = \frac{12\,V - 34.67}{20\Omega} = -1.133\,A$$

Current through a 40Ω resistance is given as:

$$i_2 = \frac{v_x}{40\Omega} = \frac{34.67}{40\Omega} = 0.866\,A$$

6. Let us find the current in a circuit using Kirchhoff's voltage law.

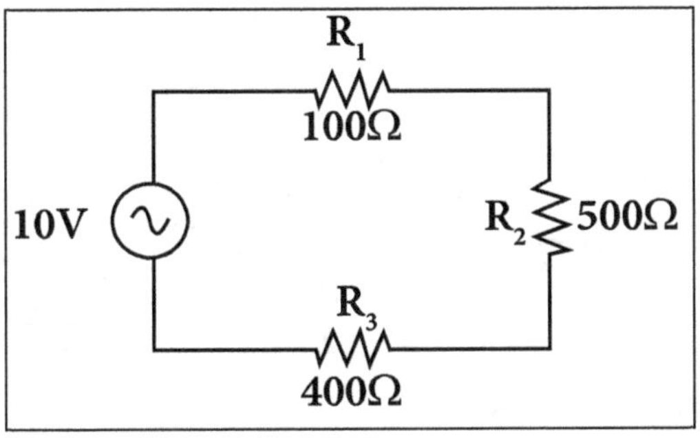

Solution:

Given:

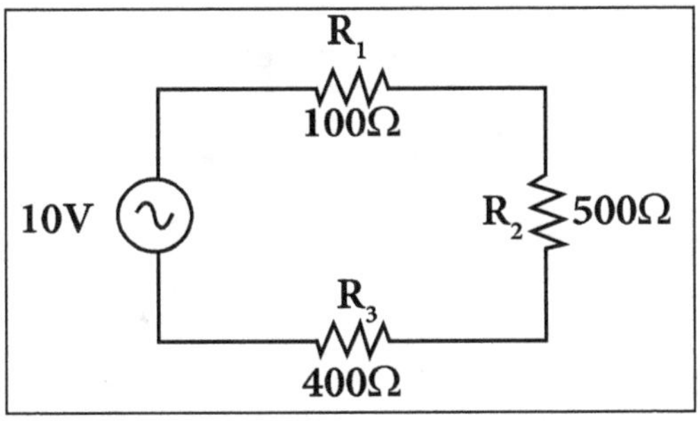

$$10 = 100(I) + 500(I) + 400(I)$$

$$10 = 1000(I)$$

$$I = 10/1000$$

$$I = 0.01A$$

7. Let us determine the current i and voltage v over each resistor.

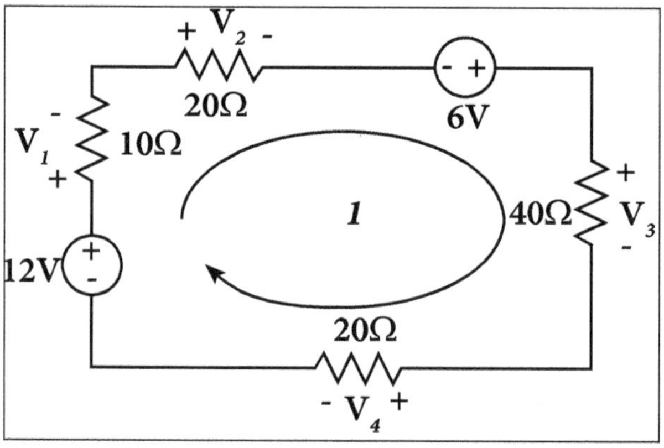

Solution:

Given:

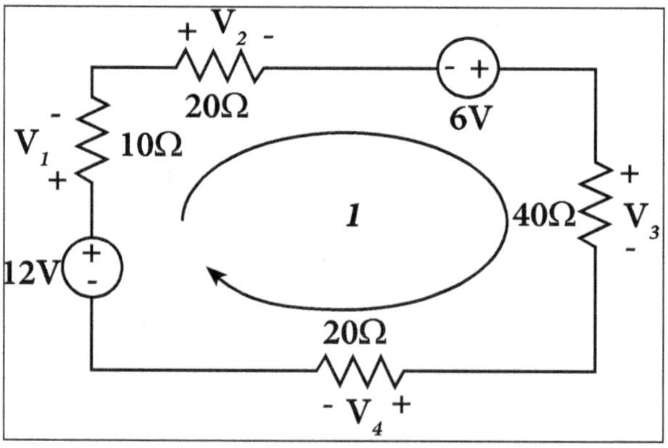

By applying KVL equations for voltages, we have:

$$V_1 + V_2 + V_3 + V_4 = 18V$$

Using ohm's law, we have:

$$R_1 = 10\Omega$$

$R_2 = 20\Omega$

$R_3 = 40\Omega$

$R_4 = 20\Omega$

Substituting into KVL equation, we get:

$$10i + 20i + 40i + 20i = 18$$

$$90i = 18$$

$$i = \frac{18}{90} = 0.2\,A$$

$$V_1 = R_1 i = 10(0.2) = 2V$$

$$V_2 = R_2 i = 20(0.2) = 4V$$

$$V_3 = R_3 i = 40(0.2) = 8V$$

$$V_4 = R_4 i = 20(0.2) = 4V$$

8. Let us determine v_1 and v_2 in the following circuit.

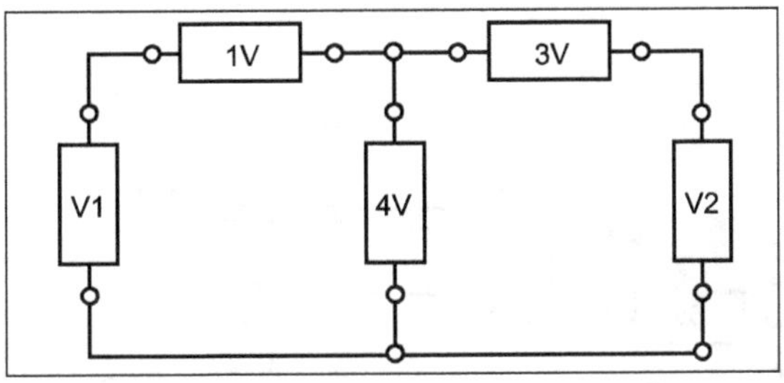

Solution:

Given:

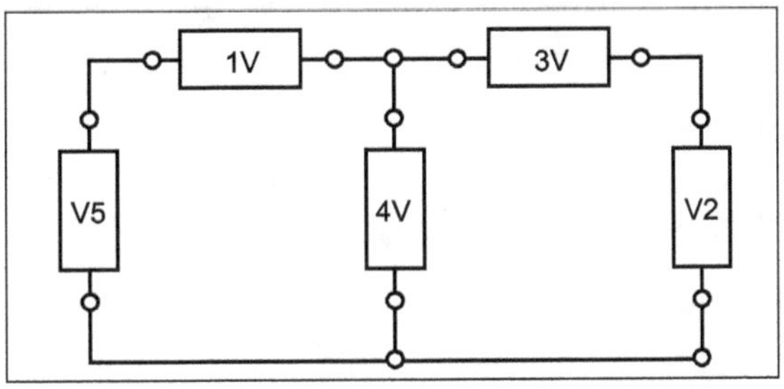

Loop 1:

$$-V_1 + 1 + 4 = 0$$

$$V_1 = 5V$$

Loop 2:

$$-4 + 3 + V_2 = 0$$

$$V_2 = 1V$$

9. Let us determine V_1, V_2 and V_3 where the arrows are signifying the positive position of the box and the negative is at the end of the box.

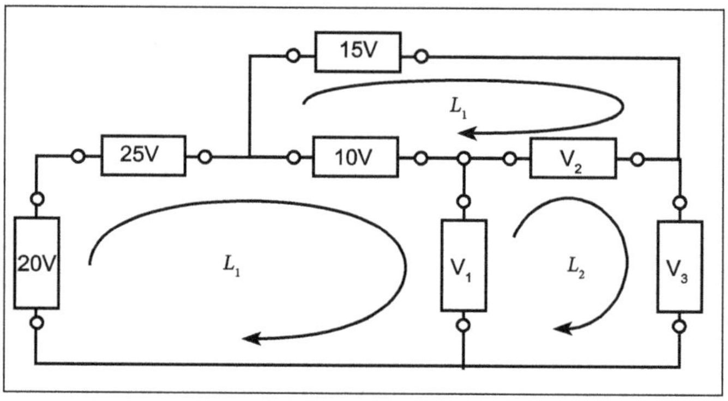

Solution:

Given:

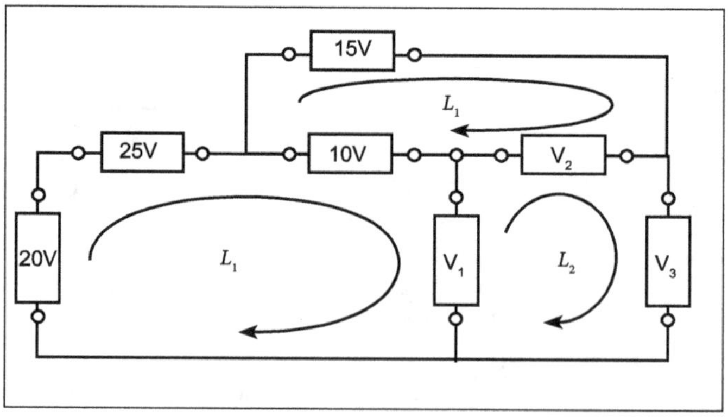

Loop 1:

$$-20 - 25 + 10 + V_1 = 0$$

$$V_1 = 35V$$

Loop 2:

$$-10+15-V_2=0$$

$$V_2=5V$$

Loop 3:

$$-V_1+V_2+V_3=0$$

$$-35+5+V_3=0$$

$$V_3=30V$$

10. Let us determine V_1, V_2, V_3 and V_4 from the below figure in which the arrows are signifying the positive position of the box and the negative is at the end of the box.

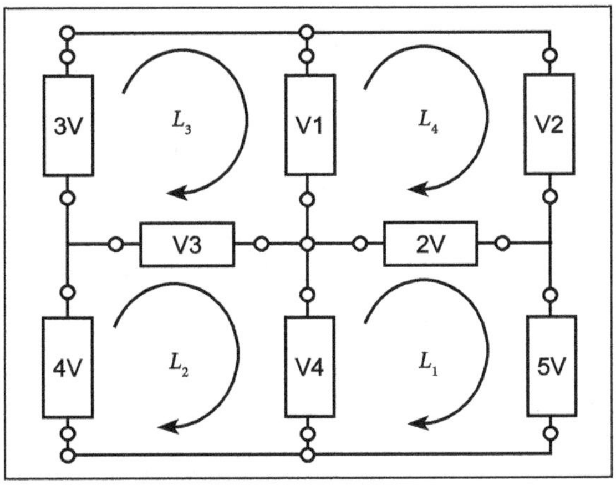

Solution:

Given:

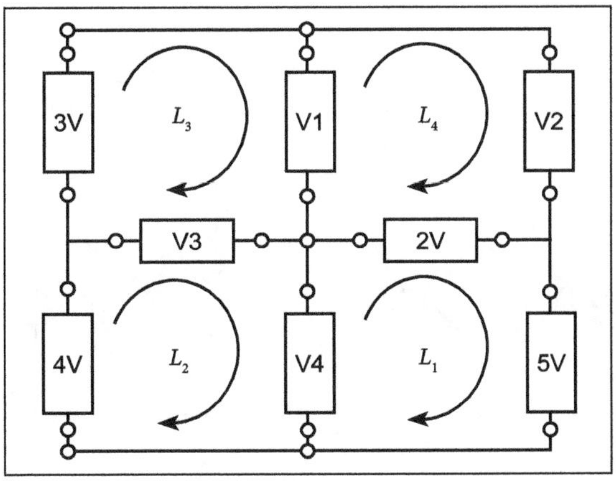

Loop 1:

$$-V_4 + 2 + 5 = 0$$

$$V_4 = 7V$$

Loop 2:

$$4 + V_3 + V_4 = 0$$

$$V_3 = -4 - 7$$

$$V_3 = -11V$$

Loop 3:

$$-3 + V_1 - V_3 = 0$$

$$V_1 = V_3 + 3$$

$$V_1 = -8V$$

Loop 4:

$$-V_1 - V_2 - 2 = 0$$

$$V_2 = -V_1 - 2$$

$$V_2 = 6V$$

11. Let us determine V_1, V_2 and V_3 in the following circuit.

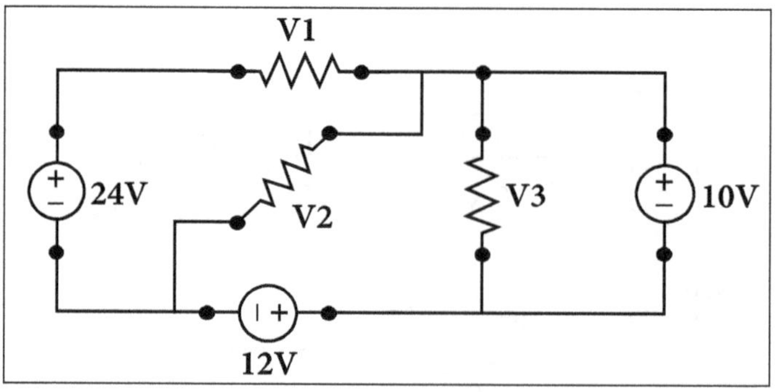

Solution:

Given:

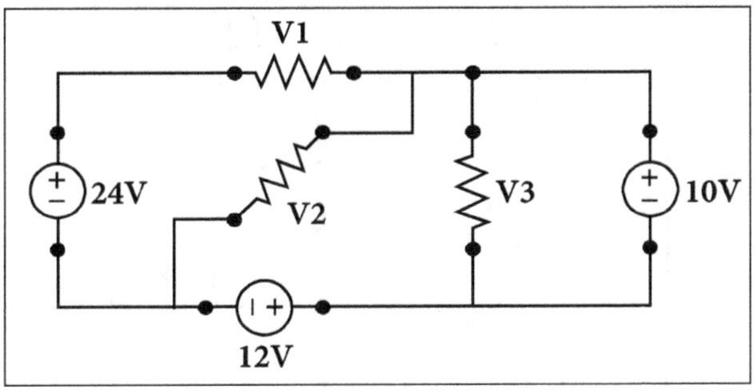

Outer loop:

$$-24+V_1+10+12=0$$

$$V_1=2V$$

V_2 and 10V source loop:

$$V_2+10+12=0$$

$$V_2=-22V$$

V_3 and 10V source loop:

$$-V_3+10=0$$

$$V_3=10V$$

12. Let us determine I_1, I_2, I_3 in the following circuit.

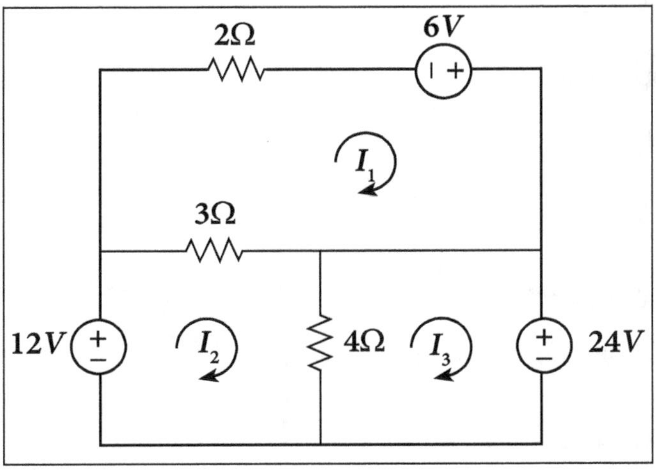

Solution:

Given:

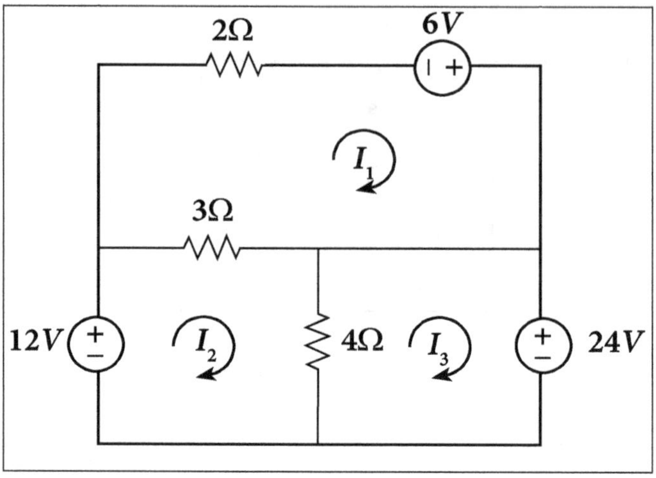

Loop 1:

$$-6 = -2\,I_1 + 3(I_2 - I_1)$$

Loop 2:

$$12 = 3(I_2 - I_1) + 4(I_2 - I_3)$$

Loop 3:

$$24 = -4(I_3 - I_2)$$

From loop 3, we have:

$$(I_2 - I_3) = 6A$$

Substituting in loop 2, we have:

$$12 = 3(I_2 - I_1) + 4(6)$$

$$I_2 - I_1 = -4A$$

Substituting in loop 1, we have:

$$-6 = -2I_1 + 3(-4)$$

$$I_1 = -3A$$

Substituting in loop 2, we have:

$$I_2 - (-3A) = -4A$$

$$I_2 = -7A$$

Substituting in loop 3, we have:

$$24 = -4\left(I_3 + 7\right)$$

$$-6 = I_3 + 7$$

$$I_3 = -13A$$

1.2 Mesh Current and Node Voltage Method of Analysis for DC and AC Circuits

Mesh Loop

Mesh is any closed path in a given circuit that does not have any element (or branch) inside it.

A mesh has the following properties:

- Every node in the closed path is exactly formed with two branches.
- No other branches are enclosed by the closed path.

Mesh Current Analysis

This is an alternative structured approach for solving the circuit which is used to calculate mesh current. A similar approach to the node situation is used.

A set of equations (based on KVL for each mesh) is formed and the equations are solved for unknown values. Many equations are needed since there is unknown mesh current.

Properties of Mesh Current Method

A mesh has the following properties:

- Every node in the closed path is exactly formed with two branches.
- No other branches are enclosed by the closed path.

Steps for analyzing the mesh current:

- Step 1: The mesh currents are identified.
- Step 2: Known mesh currents are determined.
- Step 3: Equation for each mesh is written using KVL including the mesh current and equations are solved.

Node Voltage Method

In the node voltage method, all the nodes on the circuit are identified. One of them is chosen as the reference voltage (i.e., zero potential) and other node voltages are assigned (unknown) with respect to the reference voltage.

If the circuit has 'n' nodes, then, 'n-1' node voltages are unknown. At each of these 'n-1' nodes, KCL equation is applied. The unknown node voltages are the independent variables of the problem and the solution of node voltages is obtained by solving a set of simultaneous equations.

Nodal Voltage Analysis

It determines all the node voltages in the circuit. The voltage at any given node of a circuit is the voltage drop between that node and a reference node.

All the nodes in the circuit are counted and identified. Secondly, nodes at which the voltage is already known are listed. A set of equations based on the node voltages are formed and these equations are solved for unknown quantities.

The set of equations are formed using KCL at each node. The set of simultaneous equations that are produced is then solved. Branch currents can then be found once the node voltages are known.

This can be reduced to a series of steps as follows:

Steps in Nodal Analysis

- Step 1: The nodes are identified.
- Step 2: A reference node is chosen.
- Step 3: Known node voltages are identified.
- Step 4: Branch currents are identified.
- Step 5: KCL is used to write an equation for each unknown node voltage and the equations are solved.

Problems

1. For the circuit shown in the below figure, let us find V_x using the mesh current method.

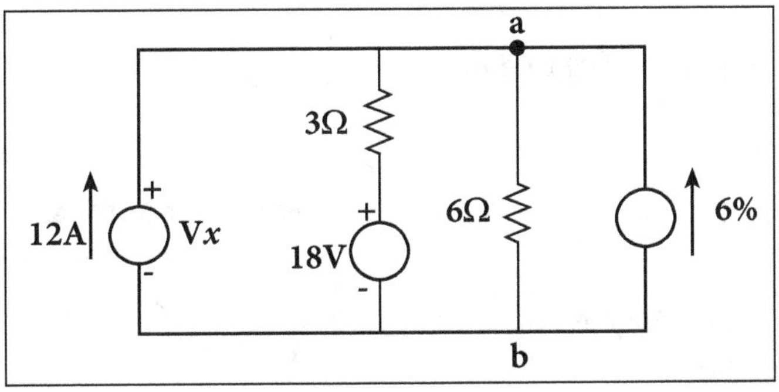

Solution:

Convert the extreme right current source (6A) into a voltage source. It is important that the current source magnitude is 6A and its internal resistance is 6Ω.

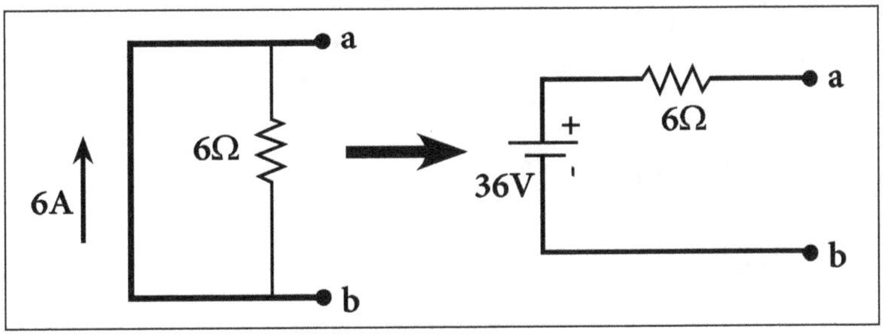

The given circuit is redrawn as shown in the below figure.

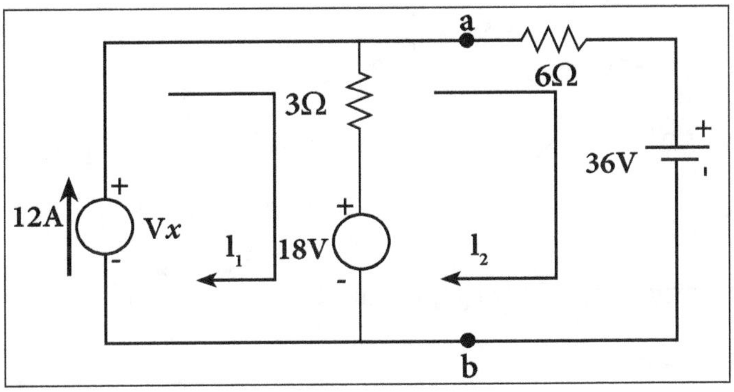

$$V_x-(I_1-I_2)\times 3-18=0 \Rightarrow V_x+3I_2=54 \qquad ...(1)$$

Where,

$I_1=12A$

Loop 2:

By using KVL, we have:

$$18-(I_2-I_1)\times 3-I_2\times 6-36=0\Rightarrow 9I_2=18\Rightarrow I_2=2A \quad ...(2)$$

By substituting the equation (2) in equation (1), we get:

$$V_x=48V$$

2. Let us find the current through branch a-b using mesh analysis.

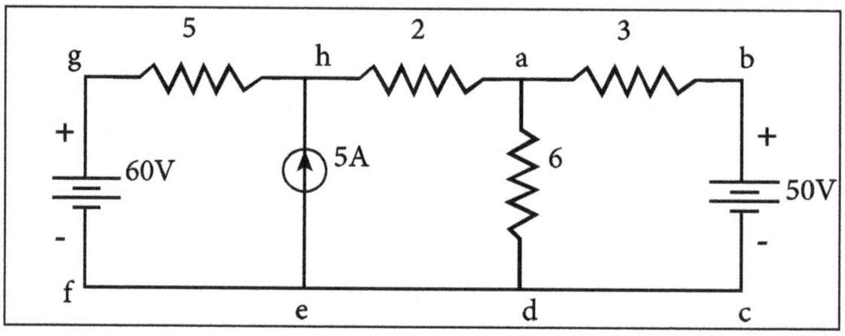

Solution:

Loop HADEH:

$$5I_1+2I_2+6(I_2-I_3)=60$$

$$5I_1+8I_2-6I_3=60 \quad ...(1)$$

Loop ABCDA:

$$3I_3+6(I_3-I_2)=-50$$

$$3I_3+6I_3-6I_2=-50$$

$$9I_3-6I_2=-50 \quad ...(2)$$

$$I_2-I_1=5A \quad ...(3)$$

From the above equation (1), (2) & (3), we have:

$$D=\begin{vmatrix}-1 & 1 & 0\\ 5 & 8 & -6\\ 0 & -6 & 9\end{vmatrix}$$

$$=-1(72-36)-1(45)$$

$$D=-81$$

$$D_3 = \begin{vmatrix} -1 & 1 & 5 \\ 5 & 8 & -60 \\ 0 & -6 & 50 \end{vmatrix}$$

$$= -1(-400+360)-(-250)+5(-30)$$

$$= 40+250-150$$

$$D_3 = 140$$

$$I_3 = D_3 / D$$

$$= 140/-81$$

$$= -1.7283$$

Current through the branch ab is 1.7283A.

3. Let us find the power in 2Ω and 5Ω resistors from the given circuit by using the nodal analysis method.

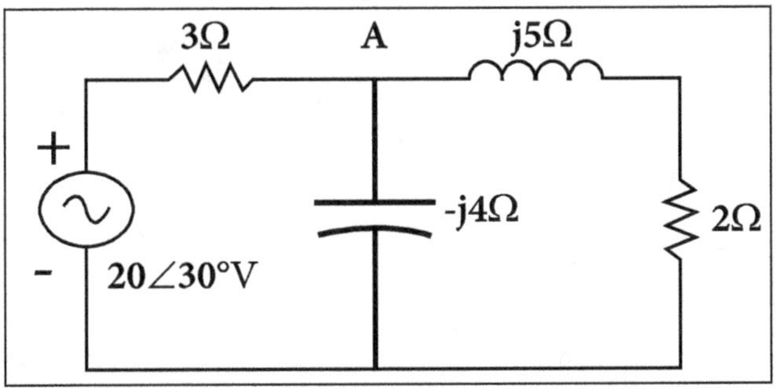

Solution:

By applying nodal analysis for the circuit shown in the above figure, power output of the source and the power in each resistor of the circuit is analyzed as follows.

Step 1:

By KCL at node A, we have:

$$\frac{V_A - 20\angle 30^\circ}{3} + \frac{V_A}{-j4} + \frac{V_A}{2+j5} = 0$$

$$V_A\left[\frac{1}{3} - \frac{1}{j4} + \frac{1}{2+j5}\right] - \frac{20\angle 30}{3} = 0$$

$$V_A[0.33+0.068+j0.077]=6.67\angle 30°$$

$$V_A[0.398+j0.077]=6.67\angle 30°$$

$$V_A=\frac{6.67\angle 30°}{[0.398+j0.078]}$$

$$V_A=\frac{6.67\angle 30°}{0.41\angle 11.08°}$$

$$V_A=16.27\angle 18.92°$$

Step 2:

Current in each resistor is given as:

$$I_{3\Omega}=\frac{V_A-20\angle 30}{3}=\frac{16.27\angle 18.92-20\angle 30}{3}=1.70\angle -112.2°$$

$$I_{2\Omega}=\frac{V_A}{2+j5}=\frac{16.27\angle 18.92}{2+j5}=3.02\angle -49.27°$$

Step 3:

Power in 3Ω resistor is given as:

$$P_{3\Omega}=I_3{}^2*3=1.70^2*3=8.67\,W$$

Step 4:

Power in 2Ω resistor is given as:

$$P_{2\Omega}=I_2{}^2*3=3.02^2*3=27.36\,W$$

4. Let us compute V_1 and V_2 in the circuit shown below using nodal analysis.

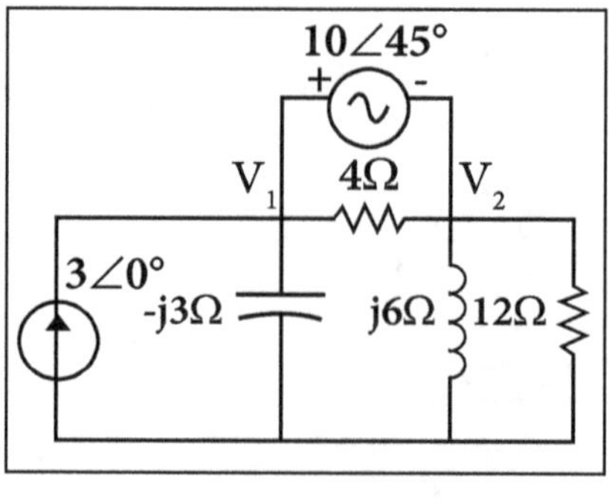

Solution:

Step 1:

Voltage difference between V_1 & V_2 is given as:

$$V_1 + V_2 = 10\angle 45°$$

$$V_1 - V_2 = 7.07 + j7.07 \quad ...(1)$$

Step 2:

By applying KCL at nodes V_1 & V_2, we have:

$$\frac{V_1}{-j3} + \frac{V_2}{j6} + \frac{V_2}{12} = 3\angle 0°$$

$$V_1(j0.33) + V_2\left(\frac{1}{j6} + \frac{1}{12}\right) = 3\angle 0°$$

$$V_1(j0.33) + V_2(0.0833 - 0.1666j) = 3 + j0 \quad ...(2)$$

Step 3:

From the above equation (1) and (2), we have:

$$\begin{bmatrix} 1 & -1 \\ j0.333 & 0.0833 - j0.1666 \end{bmatrix}\begin{bmatrix} V_1 \\ V_2 \end{bmatrix} = \begin{bmatrix} 7.07 + j7.07 \\ 3 \end{bmatrix}$$

Step 4:

V_1 and V_2 are given as:

$$V_1 = \frac{\Delta_1}{\Delta} = \frac{\begin{bmatrix} 7.07 + j7.07 & -1 \\ 3 & 0.0833 - j0.1666 \end{bmatrix}}{\begin{bmatrix} 1 & -1 \\ j0.333 & 0.0833 - j0.1666 \end{bmatrix}} = \frac{(0.0833 - j0.166) + (7.07 + j7.07) + 3}{(0.0333 j0.1666) + j0.333}$$

$$= \frac{4.7665 - j0.5886}{0.0333 + j0.1664} = 8.636 - j2.4322$$

$$V_1 = \frac{\Delta_2}{\Delta} = \frac{\begin{bmatrix} 1 & 7.07 + j7.07 \\ j0.333 & 3 \end{bmatrix}}{\begin{bmatrix} 1 & -1 \\ j0.333 & 0.0333 - j0.1666 \end{bmatrix}} = \frac{3.(7.07 + j7.07) + (j0.33)}{(0.0833 - j0.1666) + j0.333}$$

$$=\frac{5.354-j2.354}{0.0333+j0.1664}=1.567-j01.390$$

Node voltages V_1 and V_2 in polar form is given as:

$$V_1=8.636-j24.322\ (\text{or})\ 25.80\angle-70.45^\circ$$

$$V_2=1.567-j31.390\ (\text{or})\ 31.429\angle-87.14^\circ$$

1.3 Graph of a Network - Incidence and Reduced Incidence Matrices

Network Terminology

Network topology is a graphical representation of electric circuits. It is used for analyzing complex electric circuits by converting them into network graphs. Network topology is also called as graph theory.

A network graph is a network in which all nodes and loops are retained but its branches are represented by lines. The voltage sources are replaced by short circuits and current sources are replaced by open circuits. When all the elements in a network are replaced by lines with circles or dots at both ends, the configuration is called as the graph of the network.

Terminology used in Network Graph

- Path: A sequence of branches traversed from one node to another is called as a path. A set of branches can be traversed in an order without passing through the same node more than once.
- Node: A node point is defined as an end point of a line segment and exits at the junction between two branches or at the end of an isolated branch. It is a interconnection of two or more branches and it is a terminal of a branch.
- Degree of a node: It is the number of branches incident to it.
- Tree: It is an interconnected open set of branches which includes all the nodes of the given graph. In a tree of the graph, there cannot be any closed loop. A connected sub-graph contains all nodes of a graph but does not contain closed path i.e., it is a set of branches of graph which does not contain loop but connects

every node to every other node. A number of different trees can be drawn for a given graph.

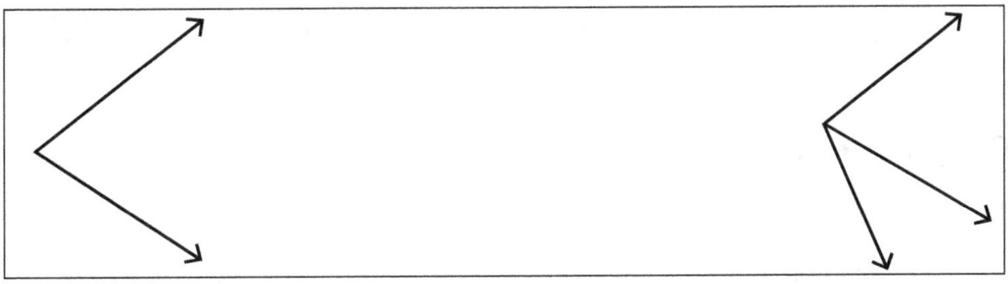

2-degree. 3-degree.

- Tree branch (Twig): It is the branch of a tree which is also named as twig.
- Tree link (Chord): It is the branch of a graph that does not belong to the particular tree.
- Loop: This is the closed contour selected in a graph.
- Cut-set: It is the set of elements or branches of a graph which separates two parts of a network. If any branch of the cut-set is not removed, then the network remains connected. It is a set of branches and if it is removed, a connected graph is divided into two connected sub graphs. Each cut set contains one tree branch and the remaining are links. The term cut-set is derived from the property which is designated by the way by which the network is divided into two parts.
- Tie-set: It is a unique set with respect to a given tree at a connected graph which contains chord and all the free branches contained in the free path are formed between two vertices of the chord. A set of branches in a loop contains one link and the remainder are tree branches.
- Network variables: A network consists of passive elements as well as sources of energy. In order to find out the response of the network, the through current and voltages across each branch of the network is obtained.
- Directed (Oriented) graph: A graph is said to be directed when all the nodes and branches are numbered or direction is assigned to the branches by arrow.
- Sub graph: A graph is said to be sub graph of a graph G if every node of it is a node of G and every branch of it is also a branch of G.
- Connected graph: When at least one path along branches between every pair of a graph exits, it is called as a connected graph.
- Incidence matrix: Any oriented graph can be described completely in a compact matrix form. The orientation of each branch in the graph and the nodes at which this branch is incident is specified. This branch is called as an incident matrix.

When one row is completely deleted from the matrix, then the remaining matrix is called as a reduced incidence matrix.

- Isomorphism: It is the property between two graphs where both have the same incidence matrix.

1.3.1 Graph of a Network

Graph

It is collection of branches and nodes in which each branch connects two nodes. A network graph is a network in which all nodes and loops are retained but its branches are represented by lines. The voltage sources are replaced by short circuits and current sources are replaced by open circuits.

The diagram that gives network geometry and uses lines with dots at the ends to represent network element is called as a graph of a given network.

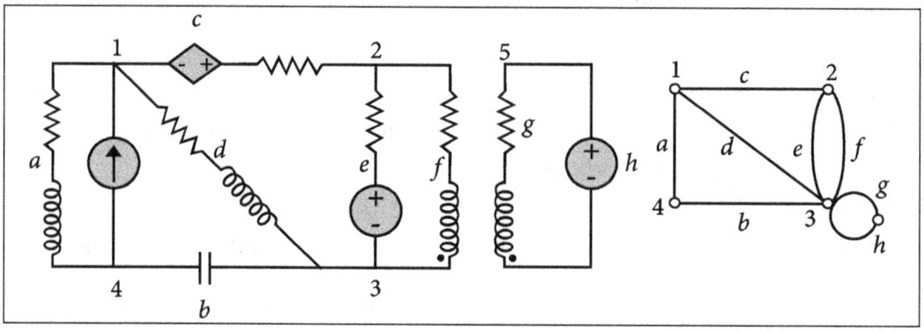

Network – graph.

Planar Graph

A graph which may be drawn on a plane surface where no branch passes over any other branch.

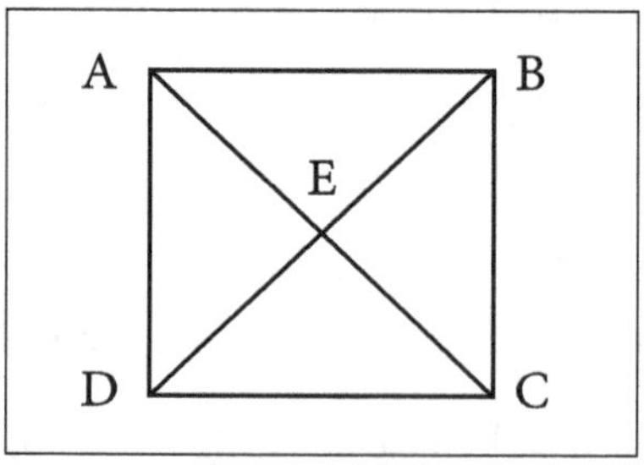

Planar graph.

Non-planar Graph

Any graph which is not planar is known as non-planar graph.

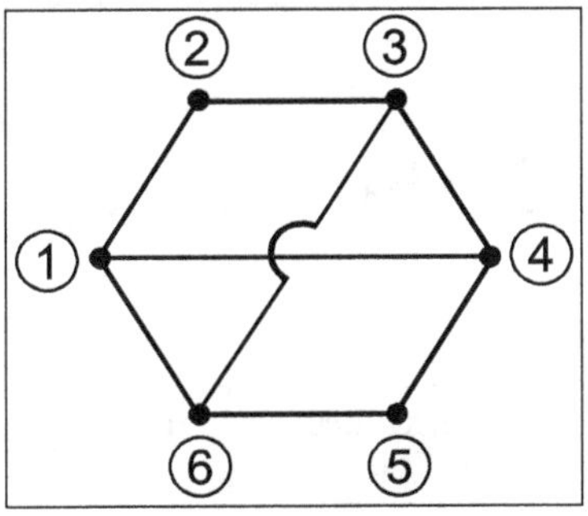

Non-planar graph.

Oriented Graph

When a direction to each branch of a graph is assigned, then the resulting graph is called as an oriented graph or a directed graph.

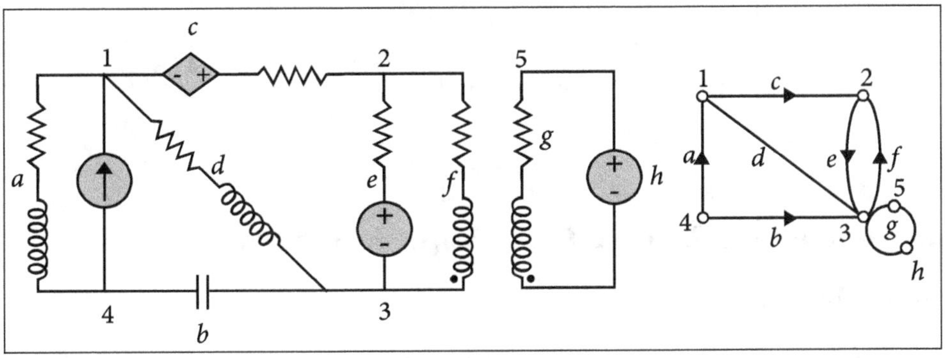

Network-oriented graph.

Connected Graph

A graph is connected if and only if there is a path between every pair of nodes.

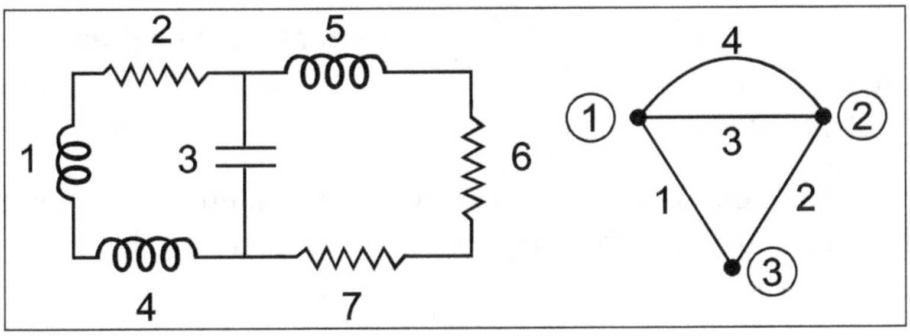

Network-connected graph.

Un-connected Graph

If there is no path from each node to every other node, then the graph is said to be un-connected graph.

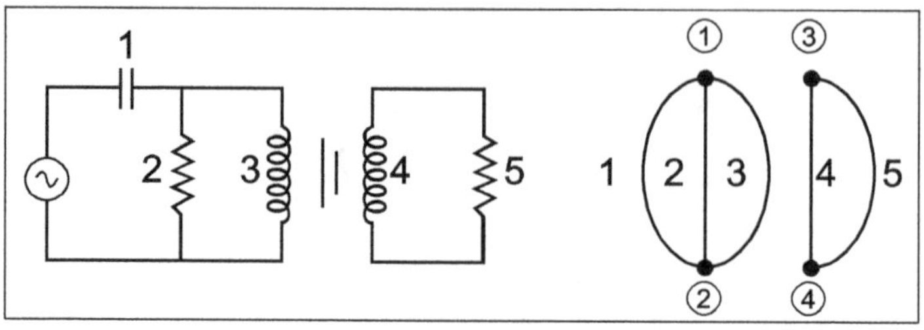

Network-unconnected graph.

Sub Graph

It is any subset of branches of the graph.

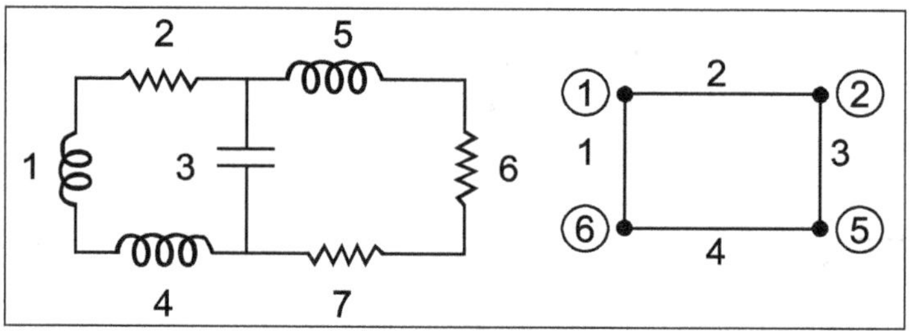

Network-sub graph.

1.3.2 Incidence and Reduced Incidence Matrices

Incidence Matrix (A_n)

Incidence matrix is a matrix which represents the graph with the help of the matrix where we can draw a graph. This matrix is denoted as $[A_C]$. In every matrix, there are rows and columns in incidence matrix $[A_C]$. The rows of the matrix $[A_C]$ represents the number of nodes and the column of the matrix $[A_C]$ represents the number of branches in the given graph.

If there are 'n' number of rows in a given incidence matrix, then there are 'n' number of nodes. If there are 'm' number of columns in a given incidence matrix, then that graph has 'm' number of branches.

It is also known as augmented incidence matrix. The element node incidence matrix 'A' indicates the incidence of elements to nodes in a connected graph. It is an N x B matrix.

$$A_n = \left(a_{kj}\right)$$

$$a_{kj} = 1$$

Branch bj is incident to and oriented away from the kth node.

$b_j = -1$ when the branch b_j is incident to and oriented towards the kth node.

$b_j = 0$ when the branch b_j is not incident to the kth node.

Since each branch of the graph is incident to exactly two nodes, we have:

$$\sum_{k=0}^{n} a_{kj} = 0 \qquad \text{for} \qquad j = 1, 2, 3, \ldots B.$$

Each column of A_n has exactly two non-zero elements where one is + 1 and the other is - 1. Sum of elements of any column is zero and the columns of A_n are lineraly dependent. The rank of the matrix is less than N.

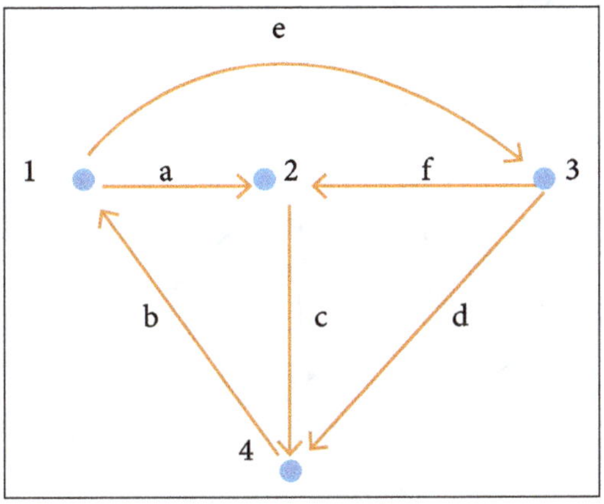

Directed graph.

In the above shown directed graph, there are 4 nodes and 6 branches. Thus, the incidence matrix for the above graph will have 4 rows and 6 columns.

The entries of incidence matrix is always -1, 0, +1. This matrix is always analogous to KCL. Thus from KCL, we can derive that,

Types of branch	**Value**
Outgoing branch from k^{th} node	+1
Incoming branch to k^{th} node	-1
Others	0

Significance of the incidence matrix translates all the geometrical features in the graph into an algebric expression.

By using the incidence matrix, we can write KCL as:

$$A_n i_B = 0$$

Where,

i_B = Branch current vector

These equations are not linearly independent. The rank of the matrix A is N - 1. This property of A_n is used to define another matrix which is called as reduced incidence matrix or bus incidence matrix.

For the oriented graph shown in the below figure the incidence matrix is as follows:

Nodes↓ branches

$$A_n = \begin{array}{c} \\ a \\ b \\ c \\ d \end{array} \begin{array}{c} \begin{matrix} 1 & 2 & 3 & 4 & 5 \end{matrix} \\ \begin{bmatrix} -1 & 1 & -1 & 0 & 0 \\ 1 & 0 & 0 & 1 & 0 \\ 0 & -1 & 0 & -1 & -1 \\ 0 & 0 & 1 & 0 & 1 \end{bmatrix} \end{array}$$

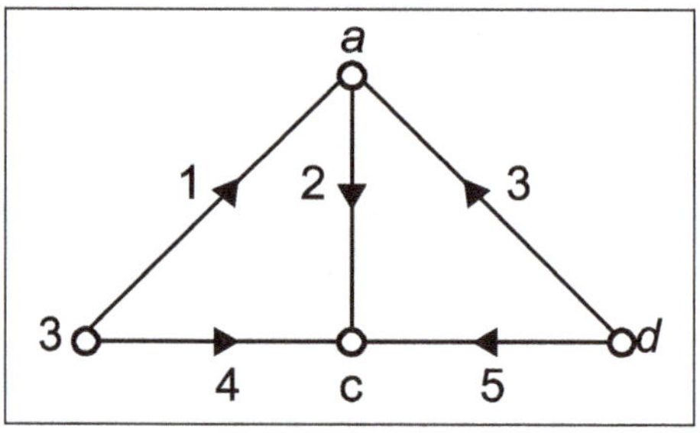

Oriented graph.

Sum of all elements in each column is zero.

Steps to Construct Incidence Matrix

Following are the steps to draw the incidence matrix:

- If a given kth node has outgoing branch, then we will write +1.
- If a given kth node has incoming branch, then we will write -1.
- Rest other branches will be considered as 0.

Example of Incidence Matrix

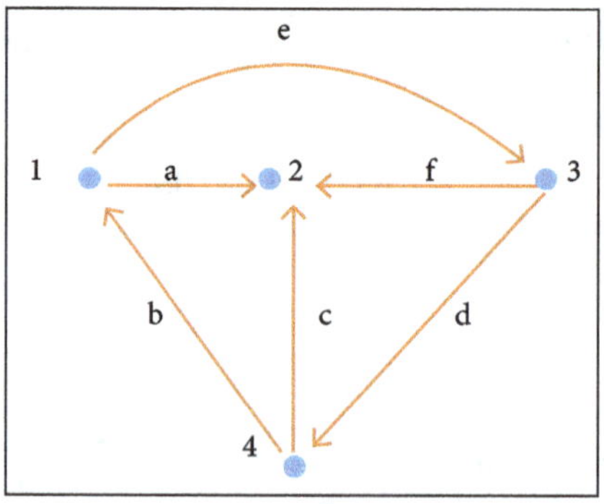

For the graph shown above, its incidence matrix is given as:

$[A_C] =$

Branches / nodes	a	b	c	d	e	f
1	1	-1	0	0	1	0
2	-1	0	-1	0	0	-1
3	0	0	0	1	1	1
4	0	1	1	-1	0	0

- If the sum of column is zero, then the incidence matrix which we have created is correct else, it is incorrect.
- For checking correctness of incidence matrix which we have drawn, we should check the sum of column.
- The incidence matrix is applied only to the directed graph only.
- The rank of complete incidence matrix is (n-1) where n is the number of nodes of the graph.
- The number of entries in a row apart from zero tells us the number of branches linked to that node. This is also called as degree of that node.
- The order of incidence matrix is (n x b) where b is the number of branches of graph.
- From a given reduced incidence matrix, a complete incidence matrix is drawn by simply adding either +1, 0 or -1 where sum of each column is zero.

Reduced Incidence Matrix

Any node of a connected graph is selected as a reference node. Then the voltages of the other nodes (referred to as buses) is measured with respect to the assigned reference.

The matrix obtained from A_n by deleting the row corresponding to the reference node is the element-bus incident matrix A and it is called as bus incidence matrix with dimension (N - 1) x B. A is rectangular and singular matrix.

From the given incidence matrix $[A_C]$, if any arbitrary row is deleted, then the new matrix will be the reduced incidence matrix. It is represented by the symbol [A]. The order of reduced incidence matrix is (n-1)*b where n is the number of nodes and b is the number of branches.

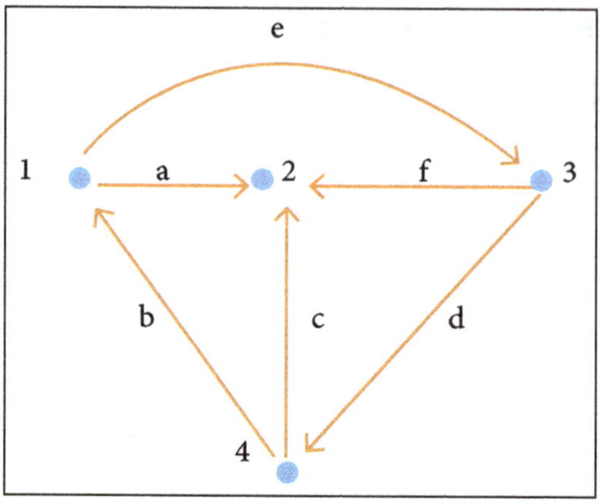

For the above shown graph, the reduced incidence matrix is given as:

[A] =

Branches / nodes	a	b	c	d	e	f
1	1	-1	0	0	1	0
2	-1	0	-1	0	0	-1
3	0	0	0	1	1	1

In the above shown matrix, row 4 is deleted.

In A_n, the sum of all elements in each column is zero. If one row is not known in A, then the sum of elements of each column must be zero.

From A, we have $A_{iB} = 0$ which represents a set of linearly independent equations and there are N - 1 independent node equations. For the graph shown in the above figure with d as the reference node, then the reduced incidence matrix is given as:

Nodes↓ branches

$$A = \begin{matrix} \\ a \\ b \\ c \end{matrix} \begin{matrix} \begin{matrix} 1 & 2 & 3 & 4 & 5 \end{matrix} \\ \begin{bmatrix} -1 & 1 & -1 & 0 & 0 \\ 1 & 0 & 0 & 1 & 0 \\ 0 & -1 & 0 & -1 & -1 \end{bmatrix} \end{matrix}$$

The sum of elements of each column in A is not zero.

Branch current vector is given as:

$$jB = \begin{bmatrix} j1 \\ j2 \\ j3 \\ j4 \\ j5 \end{bmatrix}$$

1.3.3 Trees

Tree of a connected graph is defined as any set of branches which will connect all the nodes of the graph without forming any loops. The branches of a tree are called as twigs.

A connected sub-graph contains all the nodes of a graph but there is no closed path i.e. it is a set of branches of graph which contains no loop but connects every node to every other node indirectly. A number of different trees can be drawn for a given graph.

Co-tree

Remaining branches of a graph which are not in the tree form a co-tree. The branches of a co-tree are called as links or chords.

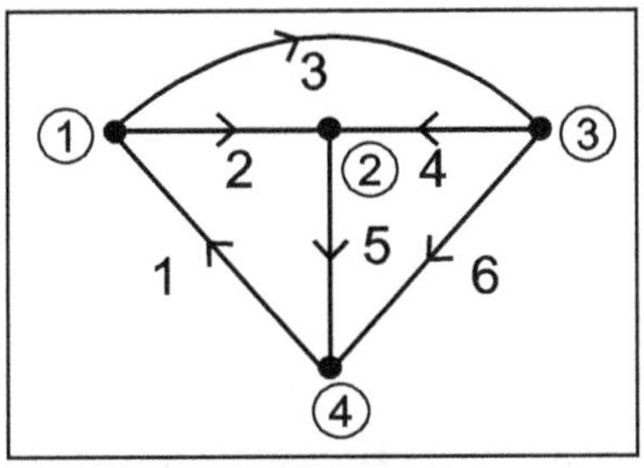

Oriented graph.

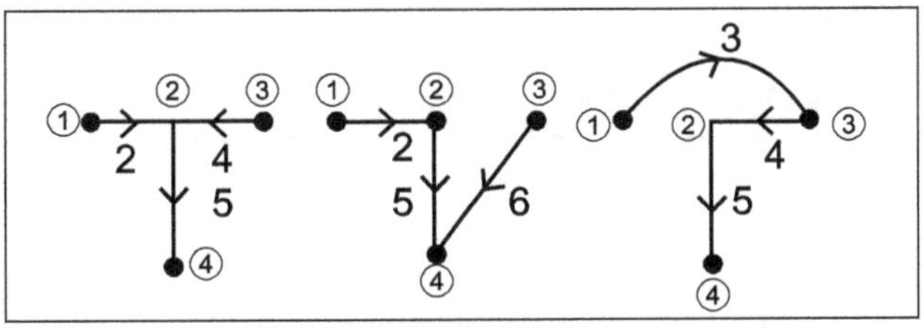

Trees.

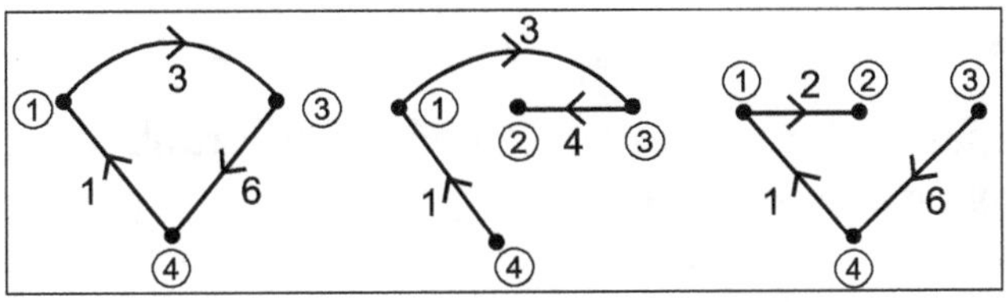

Co-trees.

Tree	Twigs	Links(Chords)
1	2, 4 and 5	1, 3 & 6
2	3, 4 and 5	1, 2 & 6
3	2, 5 and 6	1, 3 & 4

Properties of Tree

- It contains all the nodes of the graph.
- It contains (n_t - 1) branches where 'n_t' is total number of nodes in the given graph.
- There are no closed paths.

Total number of tree branches is given as:

$$n = (n_t - 1)$$

Where,

n_t = Total number of nodes.

Total number of links is given as:

$$l = (b - n)$$

Where,

b = Total number of branches in the graph.

1.4 Cut Sets and Tie Sets

Cut-sets

A cut-set of a graph is a set of branches whose removal will cut the connected graph into two parts such that the replacement of any one branch of the cut-set renders the connected two parts.

For example, two separated graphs are obtained for the graph of below figure (a) by selecting the cut-set consisting of branches [1, 2, 5, 6]. These separated graphs are shown in the below figure (b).

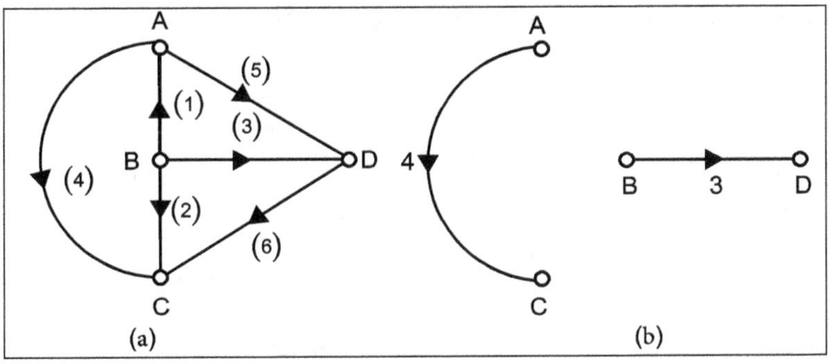

Directed graph two separate graphs created by cut set {1, 2, 5, 6}.

The cut-set is a minimal set of branches of the graph removal of which divides the graph into two connected sub-graphs. Then it separates the nodes of the graph into two groups where each being one of the two sub-graphs.

Each branch of the tie-set has one of its terminals incident at a node on one sub-graph. Selecting the orientation of the cut-set is same as that of the tree branch of the cut-set where the cut-set matrix is constructed row wise by taking one cut-set at a time.

Without link currents, the network is inactive. In the same way, without node pair voltages, the network is active. When one twig voltage is made active with all other twig voltages are zero, then the set of branches become active. This set is called as cut-set.

This set is obtained by cutting the graph by a line which cuts one twig and some links. The algebric sum of these branch currents is zero. If one twig voltage is active, then we get an entire set of node equations.

Row-by-row reading gives the KCL at each node where $QJ_B = 0$.

The procedure to write cut-set matrix is as follows:

- Oriented graph of a network is drawn and a tree is chosen.
- Each tree branch forms an independent cut-set. The direction of this cut-set is same as that of the tree branch. Each tree branch is chosen to obtain the cut-set matrix. The tree element pairs are isolated and each bridging tree branch is energized. A tree is an independent variable if the potential of bridging tree branch is equal to the node pair potential.
- Columns of the cut-set matrix is used to yield a set of equations which relates the branch potentials in terms of the node pair potentials which is obtained in matrix form as $V_B = Q^T EN$.

In the example shown in the below figure(c), (3, 4, 5) are tree branches. Links are shown in dotted lines. If two tree branch voltages in 3 and 4 are zero, then the nodes a and c are at the same potential.

Similarly, the nodes b and d are at the same potential:

$$i_5 - i_1 - i_2 - i_6 = 0$$

Since only e4 exists, the nodes a, b and c are at the same potential which reduces the graph to the form as shown in the below figure (e):

$$i_4 + i_2 + i_6 = 0$$

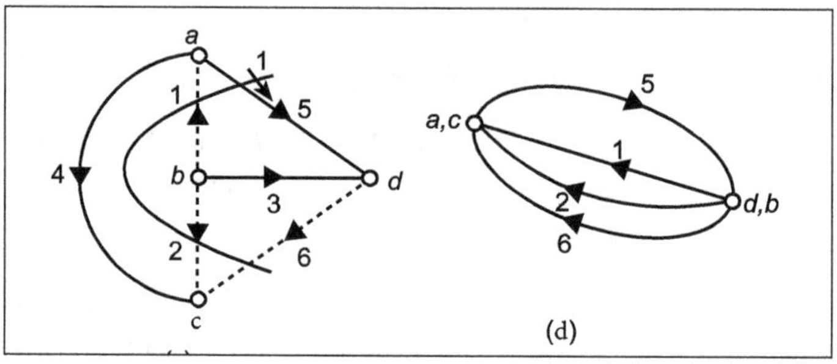

(d)

This corresponds to cut-set 2 as shown in the below figure (f):

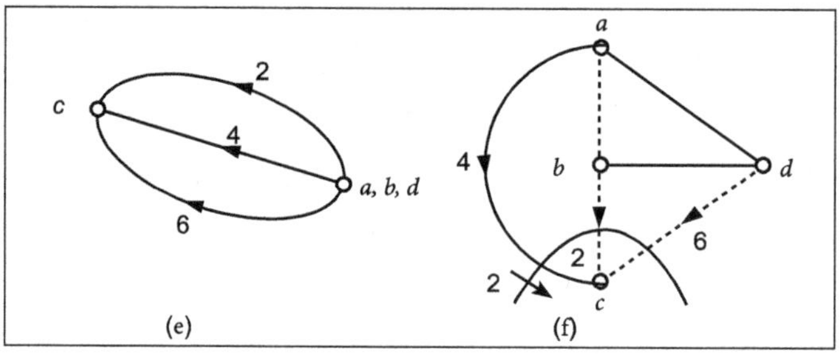

(e) (f)

For the remaining cut-set, e_4 and e_5 are made zero as in the below figure (g). e_3 is active and the nodes a, d and c are at the same potential.

Thus,

$$i_1 + i_3 + i_2 = 0$$

The corresponding cut-set 3 is shown in the below figure (h):

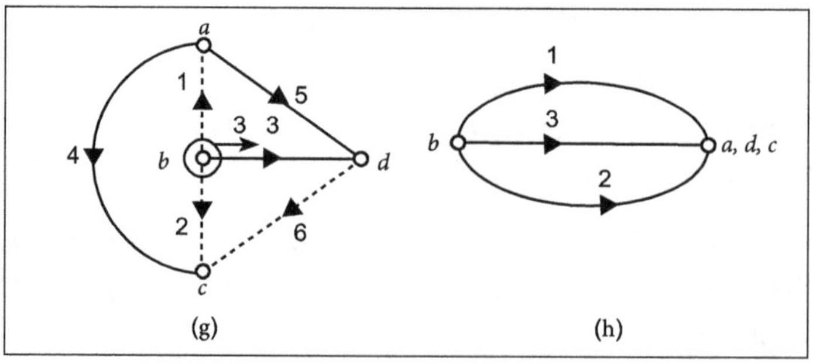

(g) (h)

The cut-set schedule is given as:

$$\begin{array}{l} \text{Node pair voltages} \downarrow \\ e_1 = v_5 \\ e_2 = v_4 \\ e_3 = v_3 \end{array} \quad \overset{\text{branches}}{\begin{array}{cccccc} 1 & 2 & 3 & 4 & 5 & 6 \end{array}} \quad \begin{bmatrix} -1 & -1 & 0 & 0 & 1 & -1 \\ 0 & 1 & 0 & 1 & 0 & 1 \\ 1 & 1 & 1 & 0 & 0 & 0 \end{bmatrix} = Q$$

QJ_B gives the following equilibrium equations:

$$-J_1 - J_2 - J_5 + J_6 = 0$$

$$J_2 + J_4 + J_6 = 0$$

$$J_1 + J_2 + J_3 = 0$$

Looking column-wise, we can express branch voltages in terms of node pair voltages as follows:

$$v_1 = -e_1 + e_3$$

$$v_2 = -e_1 + e_2 + e_3$$

$$v_3 = e_3$$

$$v_4 = e_2$$

$$v_3 = e_1$$

$$v_6 = -e_1 + e_2$$

or

$$\begin{bmatrix} -1 & 0 & 1 \\ -1 & 1 & 1 \\ 0 & 0 & 1 \\ 0 & 1 & 0 \\ 1 & 0 & 0 \\ -1 & 1 & 0 \end{bmatrix} \begin{bmatrix} e_1 \\ e_2 \\ e_3 \end{bmatrix} = V_B$$

$$V_B = Q^T E_N$$

1.4.1 Fundamental Cut Sets – Cut Set Matrix

A cut-set is a minimum set of branches of a connected graph which when removed from the graph, the graph gets separated into 2 distinct parts called as sub-graphs and the cut-set matrix is the matrix which is obtained by row-wise taking one cut-set at a time. The cut-set matrix is denoted by the symbol $[Q_f]$.

A fundamental cut set of a graph is a set of one twig and chords at a node. At each node for each twig, there will be a fundamental cut set. For a graph with n nodes, there will be n - 1 number of fundamental cut sets.

The orientation of f-cut sets is selected such that it coincides with the orientation of the defining twig off cut set.

Example:

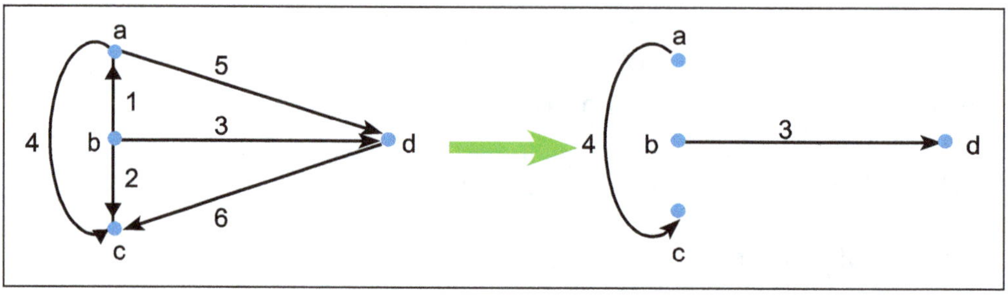

Two sub-graphs are obtained from a graph by selecting cut-sets consisting of branches [1, 2, 5, 6]. Fundamental cut set of a given graph with reference to a tree is a cut-set which is formed with one twig and remaining links.

Twigs are the branches of tree and links are the branches of co-tree. Thus, the number of cut set is equal to the number of twigs.

Number of twigs = N - 1

Where,

N = Number of nodes in the given graph.

The orientation of cut-set is the same as that of twig and it is positive.

Let us consider an oriented graph as shown in the below figure (a). One tree is selected as shown in the below figure (b). The twigs are shown by dark lines while links arc is shown by dashed lines. The fundamental cut sets are given as (n - 1) i.e., (4 - 1) = 3.

Fundamental cut sets are shown in the below figure (c), (d) and (e).

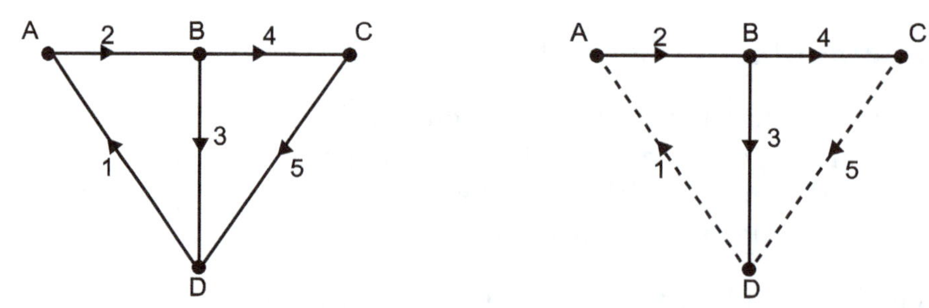

(a) Oriented graph (b) Tree

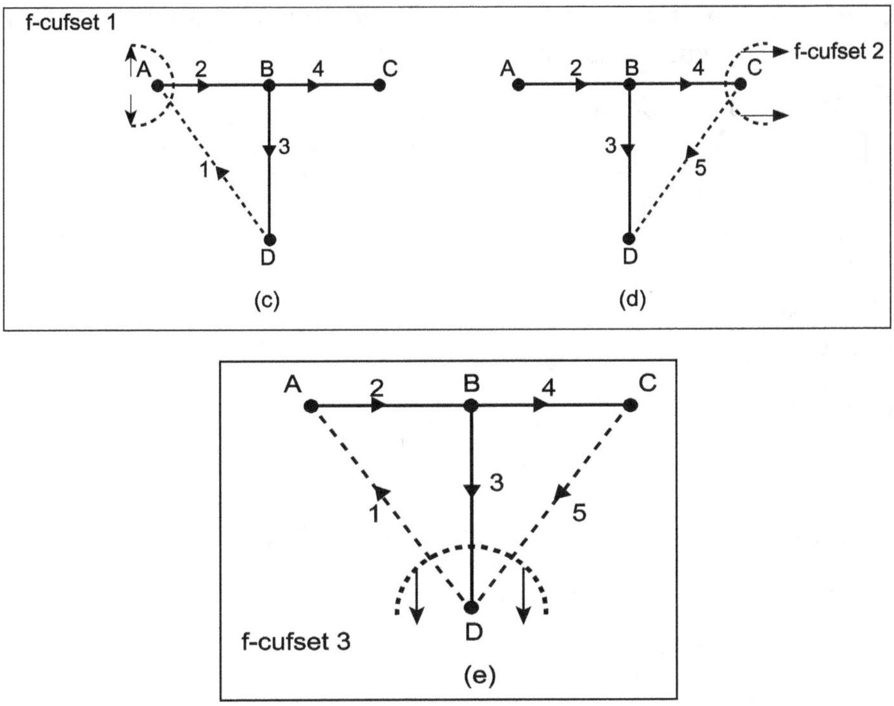

Fundamental cutsets.

The f-cut sets are given as:

$$\text{f}-\text{cutset 1}:[2,\ 1]$$

$$\text{f}-\text{cutset 2}:[4,\ 5]$$

$$\text{f}-\text{cutset 3}:[3,\ 5,\ 1]$$

The f-cut set schedule is given as follows:

f- cut set	**Branches**				
	1	2	3	4	5
f-cut set [2, 1]	-1	1	0	0	0
f-cut set [4, 5]	0	0	0	1	-1
f-cut set [3,5,1]	-1	0	1	0	1

Fundamental cut set matrix is given as:

$$\begin{matrix} & 1 & 2 & 3 & 4 & 5 \leftarrow & \text{Branches} \end{matrix}$$

$$Q=\begin{bmatrix} -1 & 1 & 0 & 0 & 0 \\ 0 & 0 & 0 & 1 & -1 \\ -1 & 0 & 1 & 0 & 1 \end{bmatrix} \begin{matrix} \text{f}-\text{cutset}1\downarrow \text{f}-\text{cutsets} \\ \text{f}-\text{cutset}2 \\ \text{f}-\text{cutset}3 \end{matrix}$$

Fundamental cut set matrix is written by writting twig entries first and then link entries. The fundamental cut set matrix is written as:

$$Q=\begin{matrix} 2 & 4 & 3 & 1 & 5 \end{matrix}$$

$$Q=\underbrace{\begin{bmatrix} 1 & 0 & 0 \\ 0 & 1 & 0 \\ 0 & 0 & 1 \end{bmatrix}}_{\text{Twigs}}\underbrace{\begin{bmatrix} -1 & 1 \\ 0 & 0 \\ -1 & 0 \end{bmatrix}}_{\text{Links}}$$

$$=[Q_T : Q_L]$$

$$=[U : Q_L]$$

Where:

Q_T = Twig matrix with all entries at the twigs of the graph.

Twig matrix is an identity matrix i.e.

$$[Q_T]=[U]=\begin{bmatrix} 1 & 0 & 0 \\ 0 & 1 & 0 \\ 0 & 0 & 1 \end{bmatrix}$$

The matrix consisting entries of links is called as link matrix $[Q_L]$.

Steps to Draw Cut Set Matrix

There are some steps one should follow while drawing the cut-set matrix. The steps are as follows:

- Graph of given network or circuit is drawn.
- Then its tree is drawn where the branches of the tree will be twig.
- Then the remaining branches of the graph are drawn by dotted line and these branches will be links.
- Each branch or twig of tree will form an independent cut-set.
- Matrix with rows is written as cut-set and column is written as branches.

Branches = 1 2 3 4 5 6.. b

Cutsets:

C_1

C_2

C_3

.

C_n

n = Number of cut-set

b = Number of branches

Orientation in Cut Set Matrix

Q_{ij} = 1 if branch J is in cut-set with orientation same as that of tree branch.

Q_{ij} = -1 if branch J is in cut-set with orientation opposite to that of branch of tree.

Q_{ij} = 0 if branch J is not in cut-set.

Problems

1. Let us draw the cut-set matrix for the following graph.

Solution:

Step 1:

The tree for the following graph is drawn.

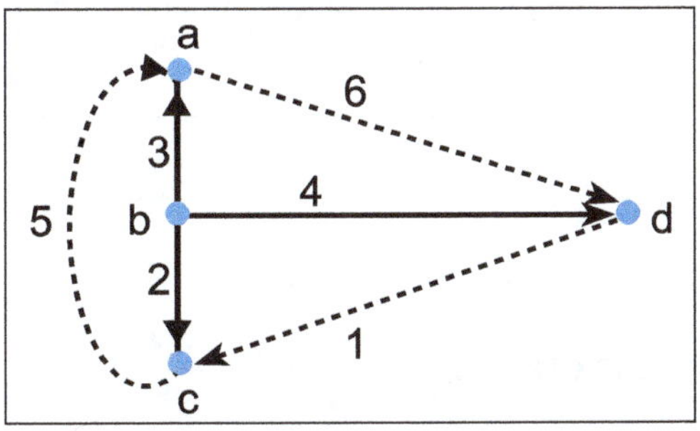

Step 2:

The cut-set is identified which is a node that contain only one twig and many links.

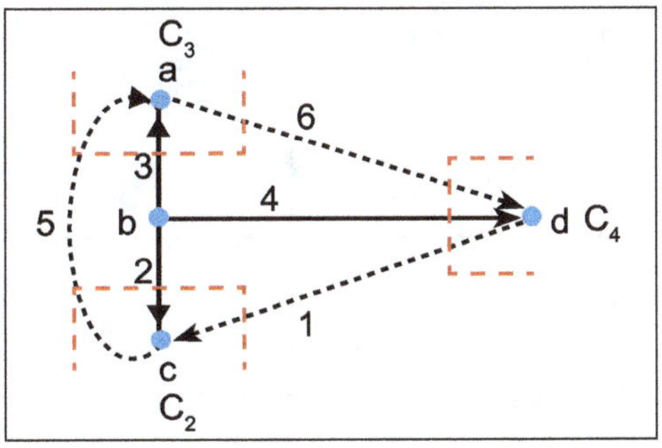

Here C_2, C_3 and C_4 are known as cut-sets.

Step 3:

The matrix is drawn.

Branches = 1 2 3 4 5 6

Cutsets:

C_2	+1	+1	0	0	−1	0
C_3	0	0	+1	0	+1	−1
C_4	−1	0	0	+1	0	+1

This is the required matrix.

2. Let us draw the cut-set of the given graph.

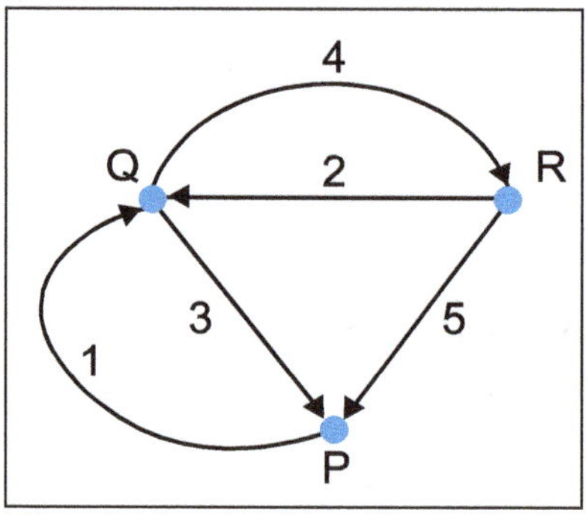

Solution:

Step 1:

Tree for the following graph is drawn.

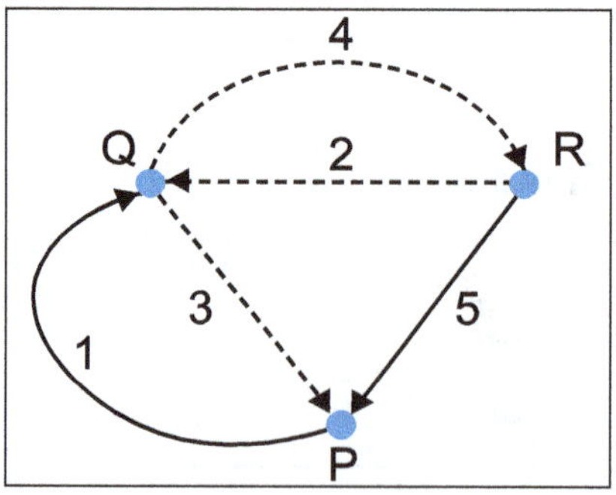

Step 2:

The cut-set is identified which contains only one twig and many links.

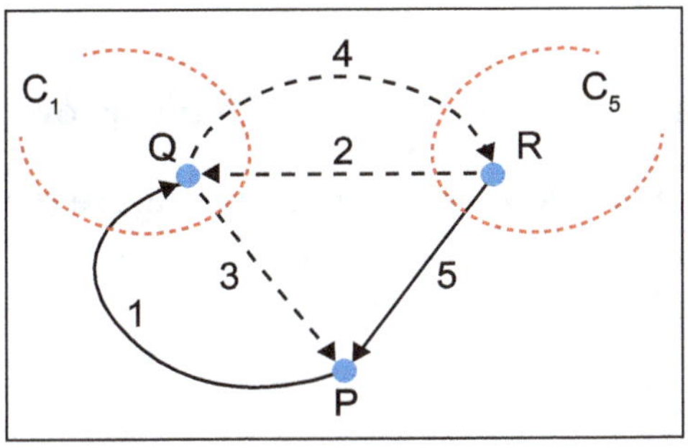

Here, C_1 and C_5 are cut-sets.

Step 3:

The matrix is drawn as shown in the below figure:

Cutsets Branches	1	2	3	4	5
C1	+1	+1	-1	-1	0
C2	0	-1	0	-1	+1

- In cut-set matrix, the orientation of twig is taken as positive.
- Each cut-set contains only one twig.
- Any number of links are attached to the cut-set.
- Relation between cut-set matrix and KCL is given as:

  $[Q_f] \times I_b = 0$

3. In the following network, the numerical values of resistances indicate the branch numbers. Let us write the oriented graph of the network. Select a tree with branches 1, 2, 3 as the tree branches and let us also write tie-set and cut-set schedule.

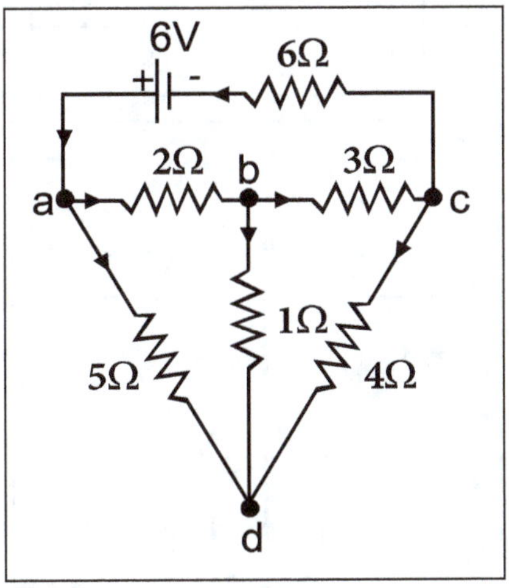

Solution:

The oriented graph of a given network is shown in the below figure (a).

The tree with branches 1, 2, 3 is shown in the below figure (b).

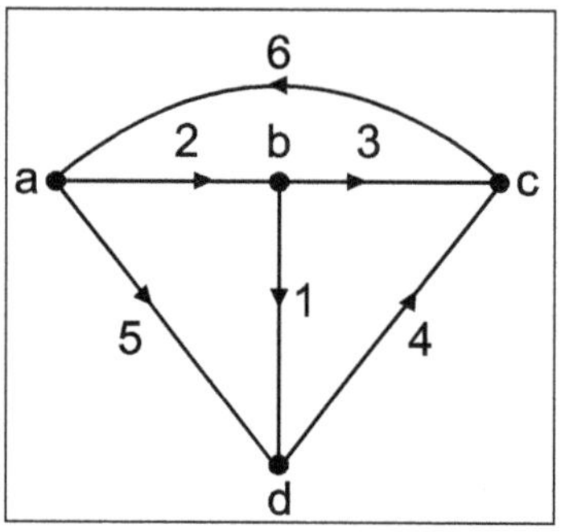

(a) Oriented graph

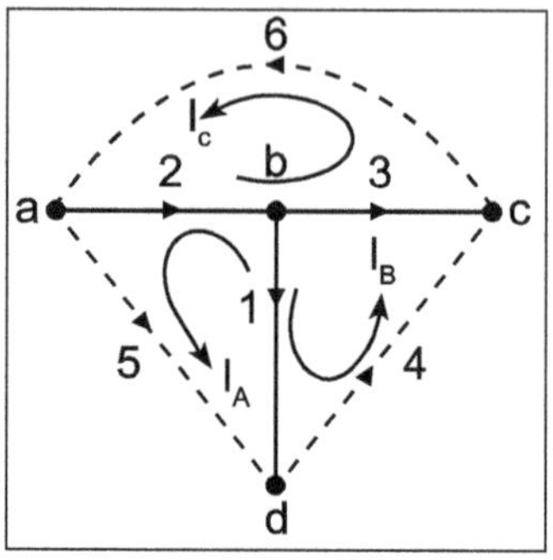

(b) Tree of a graph

The loop currents I_A, I_B, I_C are in the direction of the respective links of a tree:

The tie-sets are written as follows.

$$f-\text{tieset}\,1: I_A\left[5,1,2\right]$$
$$f-\text{tieset}\,2: I_B\left[4,3,1\right]$$
$$f-\text{tieset}\,3: I_C\left[6,2,3\right]$$

Tie-set Schedule

tiesets	Branches					
	1	2	3	4	5	6
$I_A[5, 1, 2]$	-1	-1	0	0	+1	0
$I_B[4, 3, 1]$	+1	0	-1	+1	0	0
$I_C[6, 2, 3]$	0	-1	+1	0	0	+1

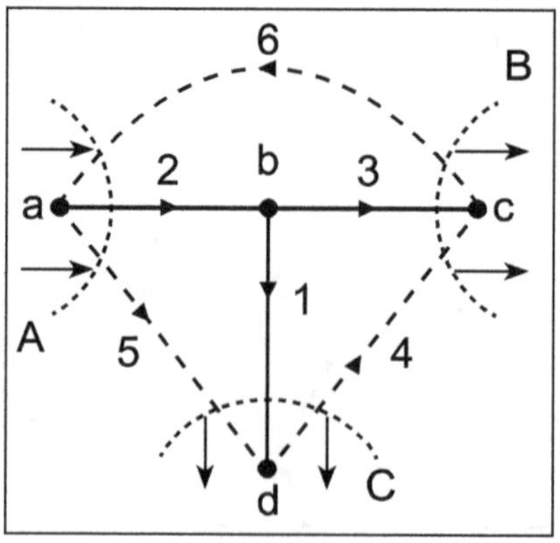

Let node 'B' is the reference node and the orientation of cutset is same as that of tree branches as shown in the above figure.

The cutsets are written as follows:

$$\text{f cutset}1\text{:}A[2, 5, 6]$$
$$\text{f cutset}2\text{:}B[3, 4, 6]$$
$$\text{f cutset}3\text{:}C[1,5, 4]$$

Cut-set schedule is written as follows:

f-cutsets	Branches					
	1	2	3	4	5	6
A[2, 5, 6]	0	+1	0	0	+1	-1
B[3, 4, 6]	0	0	+1	+1	0	-1
C[1, 5, 4]	+1	0	0	-1	+1	0

1.4.2 Tie sets

Tie-set

A tie-set is a set of branches contained in a loop such that the loop has at least one link and the remaining are twigs (Tree branches).

To construct a loop matrix, the following procedure must be followed.

Step 1:

The oriented graph of the network is drawn and a tree is chosen.

Step 2:

Each link forms an independent loop. The direction of this loop is same as that of the corresponding link and each link is chosen.

Step 3:

Tie-set matrix with elements b_{iJ} is drawn.

b_{iJ} = 1 when branch b_J is in loop i and it is directed in the same direction as the loop current.

b_{iJ} = -1 when branch b_J is in loop i and it is directed in the opposite direction as the loop current.

b_{iJ} =0 when branch b_J is not in loop i.

Tie-set matrix is an i × b matrix.

Let us consider the below example.

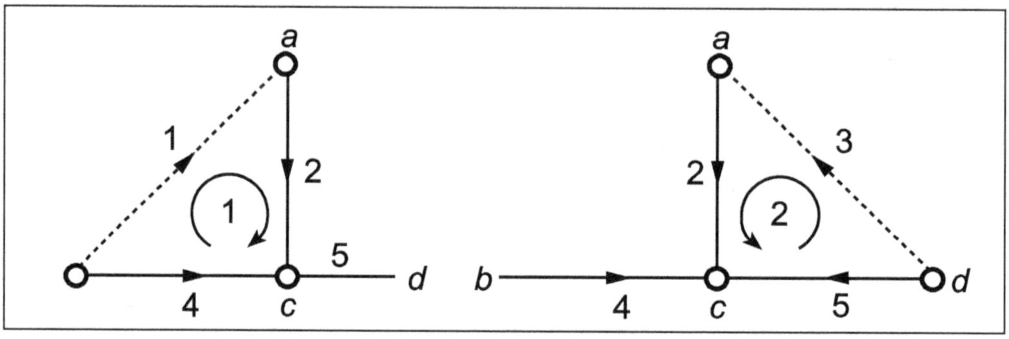

Selecting (2, 4, 5) as tree, the co-tree is (1, 3). The above figure leads to the following tie-set:

Loops ↓ branches

$$M = \begin{array}{c} \\ 1 \\ 2 \end{array}\begin{array}{c} \begin{matrix} 1 & 2 & 3 & 4 & 5 \end{matrix} \\ \begin{bmatrix} 1 & 1 & 0 & -1 & 0 \\ 0 & 1 & 1 & 0 & -1 \end{bmatrix} \end{array}$$

$$V_B = \begin{bmatrix} v1 \\ v2 \\ v3 \\ v4 \\ v5 \end{bmatrix}$$

MV_B gives the following independent loop equations as follows:

$$v_1 + v_2 - v_4 = 0$$

$$v_2 + v_3 - v_5 = 0$$

Looking column wise, we can express branch currents in terms of loop currents. This is done by the following matrix equation:

$$J_B = M^T I_L$$

The above matrix equation gives:

$$J_1 = i_1$$

$$J_2 = i_1 + i_2$$

$$J_3 = i_2$$

$$J_4 = -i_1 - i_2$$

J stands for branch current while i stand for loop current.

In this matrix:

- Each row corresponds to an independent loop. Therefore the columns of the resulting schedule yield a set of equations which relates each branch current to the loop currents.
- As each column expresses a branch current in terms of loop currents, the rows of the matrix automatically yield the closed paths in which the associated loop currents circulate.

Expressions for branch currents in terms of loop currents are obtained in matrix form as:

$$J_B = M^T I_L.I$$

Where:

M = Tie-set matrix of L × B

$$J_B = \begin{bmatrix} J_1 \\ J_2 \\ J_3 \\ J_4 \end{bmatrix}$$

$$I_L = \begin{bmatrix} i_1 \\ i_2 \end{bmatrix}$$

1.4.3 Twig Voltages and Cutset Schedules

Twigs and Links

The branches of a tree are called as twigs. For a given graph, the complementary set of branches of the tree is called as the co-tree of the graph. The branches of a co-tree are called as links i.e., those elements of the connected graph that are not included in the tree links forms a sub graph. For example, the set of branches (b, d, f) represented by dotted lines in the below figure form a co-tree of the graph.

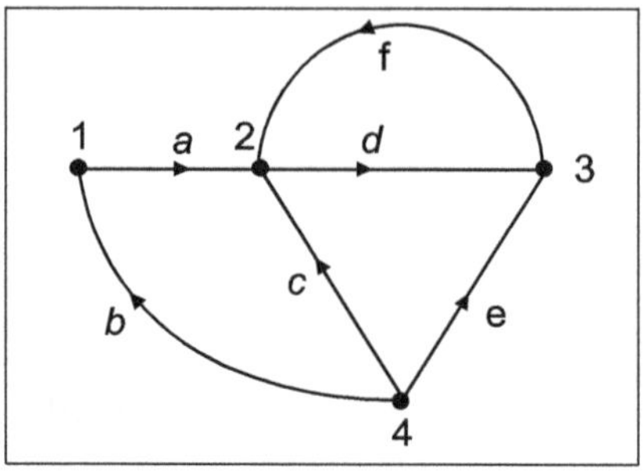

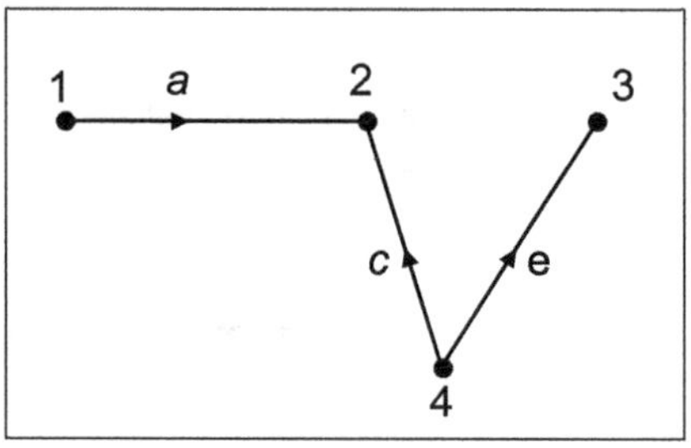

Tree.

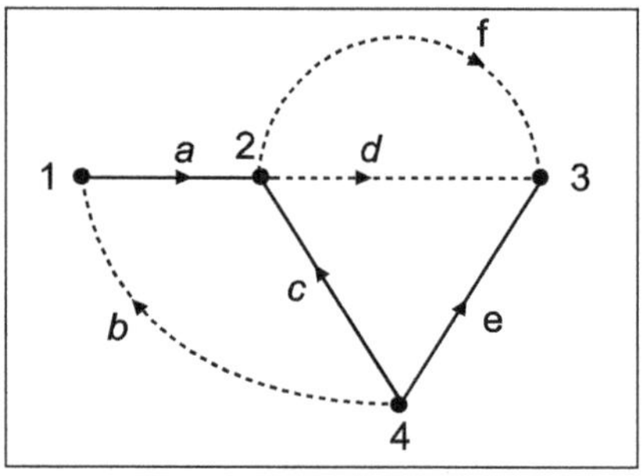

Co-tree.

The branches a, c and e are the twigs while the branches b, d and f are the links of this tree. For a network with b branches and n nodes, the number of twigs for a selected tree is (n - 1) and the number of links I with respect to this tree is (b - n + 1). The number of twigs (n - 1) is known as the tree value of the graph.

It is also called as the rank of the tree. If a link is added to the tree, the resulting graph contains one closed path called as a loop. The addition of each subsequent link forms one or more additional loops. Loops which contain only one link are independent and they are called as basic loops.

Tie-set

It is a set of branches in a loop where each loop contains one link or chord and the remainder are tree branches.

Let us consider the graph and the tree as shown in the below figure. This selected tree will result in three fundamental loops since each link is connected to the tree.

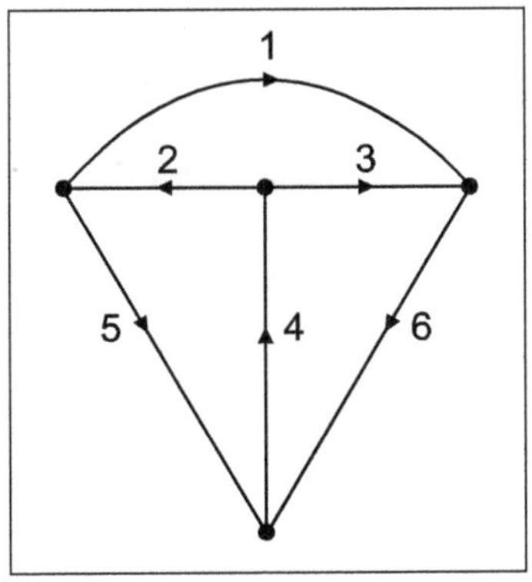

(a) Graph.

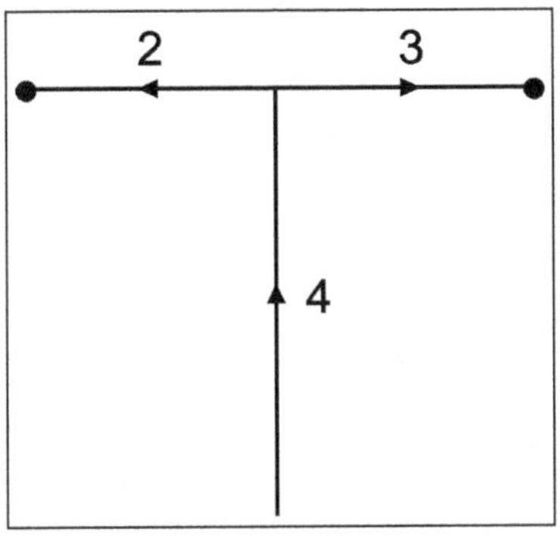

(b) Tree of the graph.

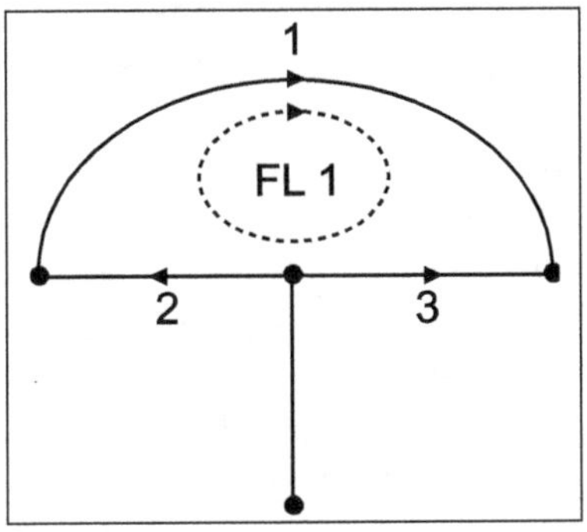

(c) Loop-1.

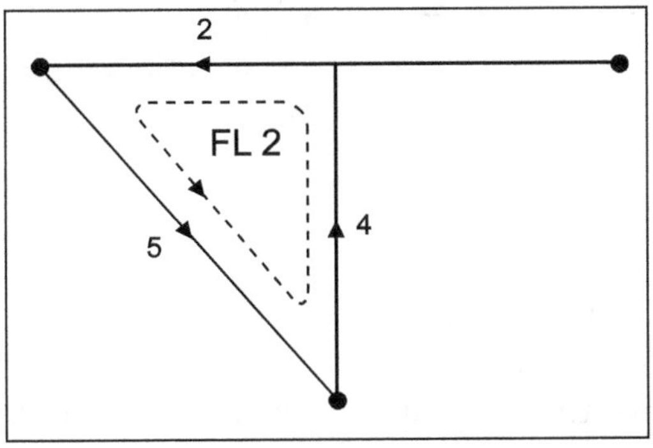

(d) Loop-2.

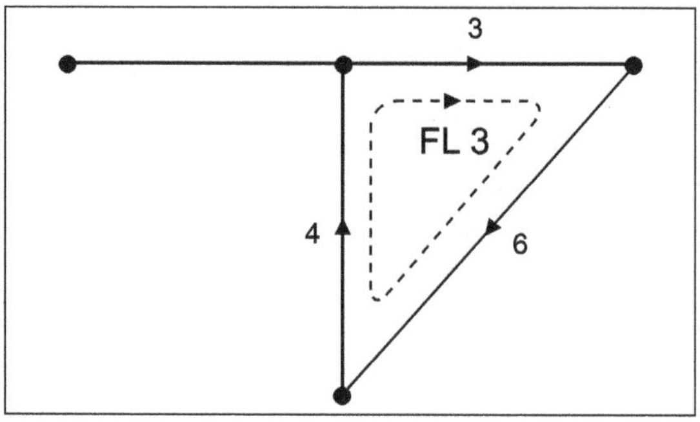

(e) Loop-3.

Fundamental loop 1(FL_1):

Link 1 is connected to the tree.

Fundamental loop 2 (FL_2):

Link 5 is connected to the tree.

Fundamental loop 3 (FL_3):

Link 6 is connected to the tree.

These sets of branches (1, 2, 3), (2, 4, 5) and (3, 4, 6) form three tie-sets.

For a given network tree, a systematic way of indicating the links through the use of a schedule is called as tie-set schedule.

The tie-set schedule for the given network oriented graph is shown below:

Link Current or Number	**Branches**							
	1	2	3	4	5	6	7	8
1	1	0	0	0	-1	0	0	1
2	0	1	0	0	1	-1	0	0
3	0	0	1	0	0	1	-1	0
4	0	0	0	1	0	0	1	-1

Tie-set schedule written in matrix form is known as tie-set matrix (B):

$$B=\begin{pmatrix} 1 & 0 & 0 & 0 & -1 & 0 & 0 & 1 \\ 0 & 1 & 0 & 0 & 1 & -1 & 0 & 0 \\ 0 & 0 & 1 & 0 & 0 & 1 & -1 & 0 \\ 0 & 0 & 0 & 1 & 0 & 0 & 1 & -1 \end{pmatrix}$$

After writing the schedule, the columns of a schedule or matrix gives branch currents in terms of link currents. Thus, column 1 gives equation for j_1 in terms of link currents:

$$\left.\begin{matrix} j_1 = i_1 & j_5 = i_2 - i_1 \\ j_2 = i_2 & j_6 = i_3 - i_2 \\ j_3 = i_3 & j_7 = i_4 - i_3 \\ j_4 = i_4 & j_8 = i_1 - i_4 \end{matrix}\right\} \quad ...(1)$$

The rows of the schedule gives KVL equations in terms of coefficients of the schedule or matrix:

$$\left.\begin{matrix} e_1 - e_5 + e_8 = 0 \\ e_2 + e_5 - e_6 = 0 \\ e_3 + e_6 - e_7 = 0 \\ e_4 + e_7 - e_8 = 0 \end{matrix}\right\} \quad ...(2)$$

The set of the above equation (1) can be expressed in the matrix form as follows:

$$\begin{pmatrix} j_1 \\ j_2 \\ j_3 \\ j_4 \\ j_5 \\ j_6 \\ j_7 \\ j_8 \end{pmatrix} = \begin{pmatrix} 1 & 0 & 0 & 0 \\ 0 & 1 & 0 & 0 \\ 0 & 0 & 1 & 0 \\ 0 & 0 & 0 & 1 \\ -1 & 1 & 0 & 0 \\ 0 & -1 & 1 & 0 \\ 0 & 0 & -1 & 1 \\ 1 & 0 & 0 & -1 \end{pmatrix} \begin{pmatrix} i_1 \\ i_2 \\ i_3 \\ i_4 \end{pmatrix} \quad ...(3)$$

In compact form, we have:

$$(I_b) = (B)^T (I_1) \quad ...(4)$$

Where:

(I_b) = Column matrix of branch currents of the order (b x 1)

$(B)^T$ = Transpose of the fundamental tie-set matrix B

(I_1) = Column matrix of loop currents or link currents of the order (1 x 1)

Where:

'I' = Number of independent loops

From the set of equation (2), we have:

$$\begin{pmatrix} 1 & 0 & 0 & 0 & -1 & 0 & 0 & 1 \\ 0 & 1 & 0 & 0 & 1 & -1 & 0 & 0 \\ 0 & 0 & 1 & 0 & 0 & 1 & -1 & 0 \\ 0 & 0 & 0 & 1 & 0 & 0 & 1 & -1 \end{pmatrix} \begin{pmatrix} e_1 \\ e_2 \\ e_3 \\ e_4 \\ e_5 \\ e_6 \\ e_7 \\ e_8 \end{pmatrix} = 0 \quad ...(5)$$

In compact form, we have:

$$(B)(E_b) = 0 \quad ...(6)$$

(E_b)= Column matrix of branch voltages of the order (b x 1)

The above equations (4) and (6) are the basic equations required to derive equilibrium equations on loop current basis.

Cut-set Schedule

For a given network tree, a systematic way of indicating the tree branch voltage through the use of a schedule is called as a cut-set schedule.

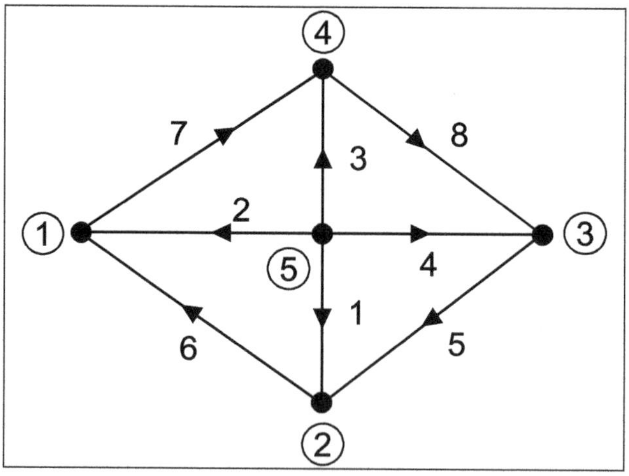

Oriented graph.

Steps to write the cut-set schedule for network graph:

- Let us consider an oriented network graph.
- Any one possible tree of the network graph is written.

- Let us assume tree branch voltages as (e1, e2...en) independent variables.
- Let us also assume that the independent voltage variable is in same direction as that of a tree branch voltage.
- The cut-sets are marked in the network graph.

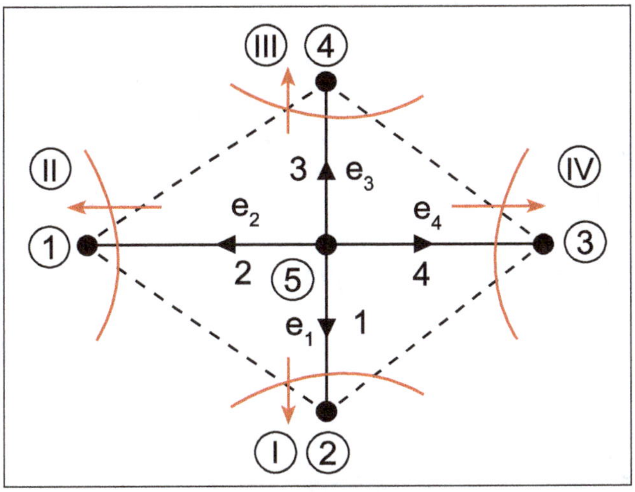

Tree and cut-sets.

The tree branch voltages e_1, e_2, e_3, & e_4 entered in the first column of the schedule correspond to 4 branches 1, 2, 3 & 4. In order to fill the first row corresponding to the tree branch voltages el, the direction of currents in the branches connected to the cut-set is considered.

If the direction of current in the cut-set branch is towards the cut-set node, '+1' is written in the branch column of the concerned cut-set branch. If the direction of current in the cut-set branch is away from the cut-set node, then '-1' is written in that particular cut-set branch column. '0' is written in the branch columns which are not in that particular cut-set.

Cut-set Schedule:

Tree Branches Voltages	**Branches**							
	1	2	3	4	5	6	7	8
e_1	1	0	0	0	1	-1	0	0
e_2	0	1	0	0	0	1	-1	0
e_3	0	0	1	0	0	0	1	-1
e_4	0	0	0	1	-1	0	0	1

The columns of the cut-set schedule gives branch voltage equations as follows.

$$\left.\begin{aligned} v_1 &= e_1 \\ v_2 &= e_2 \\ v_3 &= e_3 \\ v_4 &= e_4 \\ v_5 &= (e_1 - e_4) \\ v_6 &= (-e_1 + e_2) \\ v_7 &= (-e_2 + e_3) \\ v_8 &= (-e_3 + e_4) \end{aligned}\right\} \qquad \ldots(1)$$

The rows of cut-set schedule gives KCL equations.

$$\left.\begin{aligned} j_1 + j_5 - j_6 &= 0 \\ j_2 + j_6 - j_7 &= 0 \\ j_3 + j_7 - j_8 &= 0 \\ j_4 - j_5 - j_8 &= 0 \end{aligned}\right\} \qquad \ldots(2)$$

The cut-set schedule written in the matrix form is known as cut-set matrix which is represented as Q.

$$Q = \begin{pmatrix} 1 & 0 & 0 & 0 & 1 & -1 & 0 & 0 \\ 0 & 1 & 0 & 0 & 0 & 1 & -1 & 0 \\ 0 & 0 & 1 & 0 & 0 & 0 & 1 & -1 \\ 0 & 0 & 0 & 1 & -1 & 0 & 0 & 1 \end{pmatrix}$$

From the set of equations (1), we have:

$$\begin{pmatrix} v_1 \\ v_2 \\ v_3 \\ v_4 \\ v_5 \\ v_6 \\ v_7 \\ v_8 \end{pmatrix} = \begin{pmatrix} 1 & 0 & 0 & 0 \\ 0 & 1 & 0 & 0 \\ 0 & 0 & 1 & 0 \\ 0 & 0 & 0 & 1 \\ 1 & 0 & 0 & -1 \\ -1 & 1 & 0 & 0 \\ 0 & -1 & 1 & 0 \\ 0 & 0 & -1 & 1 \end{pmatrix} \begin{pmatrix} e_1 \\ e_2 \\ e_3 \\ e_4 \end{pmatrix}$$

In compact form, we have:

$$[V_b] = [Q]^T [V_T] \qquad \ldots(3)$$

Where:

$[V_b]$ = Branch voltage column matrix of order b x 1

$[V_T]$ = Tree branch voltage column matrix of order n x 1

$[Q]$ = Cut-set matrix of order n x b

$$\begin{pmatrix} 1 & 0 & 0 & 0 & 1 & -1 & 0 & 0 \\ 0 & 1 & 0 & 0 & 0 & 1 & -1 & 0 \\ 0 & 0 & 1 & 0 & 0 & 0 & 1 & -1 \\ 0 & 0 & 0 & 1 & -1 & 0 & 0 & 1 \end{pmatrix} \begin{pmatrix} j_1 \\ j_2 \\ j_3 \\ j_4 \\ j_5 \\ j_6 \\ j_7 \\ j_8 \end{pmatrix} = 0$$

In compact form, we have:

$$[Q][I_b]=0 \qquad \text{...(4)}$$

Where:

$[I_b]$ = Branch current column matrix of order b x 1

$[Q]$ = Cut-set matrix of order n x b

1.5 Duality and Dual Networks

Duality

Duality is a mutual relationship. A number of similarities in analyzing the network depends on current (Loop) basis and voltage (Node) basis. The principal quantities involved in the two methods form pairs. Each of the quantity in such a pair plays a dual role and the quantities which form pair are called as dual quantities.

Principle of Duality: Identical behavior patterns observed between voltages and currents between two independent circuits illustrate the principle of duality.

Let us consider a network containing R, L & C elements connected in series and excited by a voltage source as shown in the below figure.

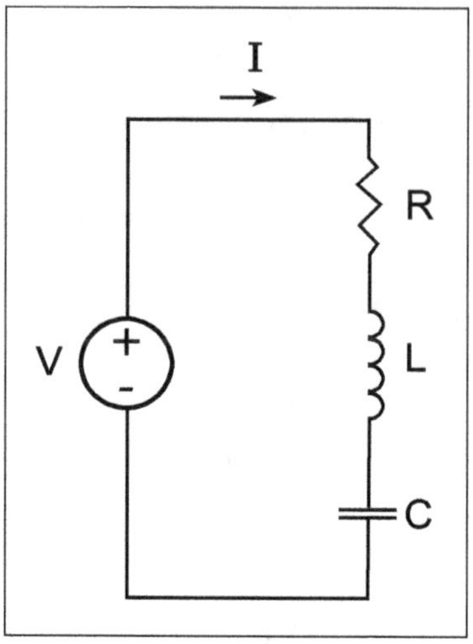

Series RLC circuit.

Integral differential equations for the above circuit are given as:

$$Ri + L\,(di/dt) + (1/C)\int i\,dt = v \quad ...(1)$$

Let us consider a network containing R, L & C elements connected in parallel and driven by a current source as shown in the below figure.

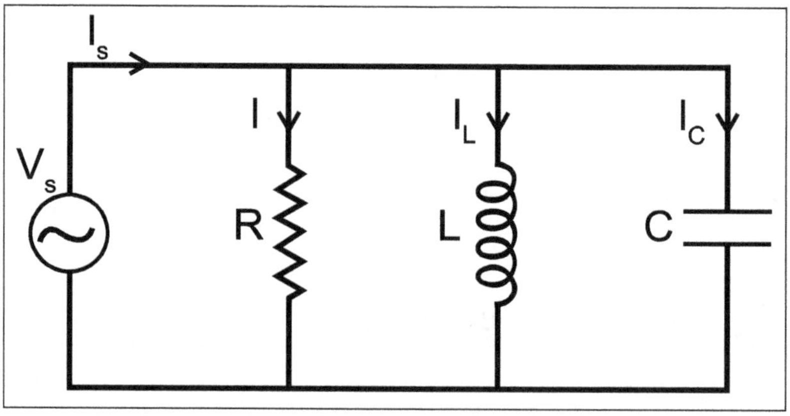

Parallel RLC circuit.

The integral differential equations for the above circuit are given as:

$$(1/R)\,v + C\,(dv/dt) + (1/L)\int v\,dt = i \quad ...(2)$$

$$G\,v + C\,(dv/dt) + (1/L)\int v\,dt = i$$

Both the equations have same solutions.

Therefore, these two networks are called as duals.

Construction of a Dual Network

Only a planar networks without mutual inductances have duals.

S. No.	Quantity	Dual quantity
1.	Current	Voltage
2.	Resistance	Conductance
3.	Inductance	Capacitance
4.	Branch current	Branch voltage
5.	Mesh	Node
6.	Loop	Node-pair
7.	Number of loops	Number of node-pairs
8.	Loop current	Node-pair voltage
9.	Mesh current	Node voltage or node potential
10.	Link	Tree branch
11.	Tie-set	Cut-set
12.	Short circuit	Open circuit
13.	Parallel path	Series path
14.	Charge(Q)	Flux linkages(ψ)

Procedure to Draw a Dual Network

These rules illustrated below are only for planar or flat networks which do not have any of their branches crossing other branches 1.

- Place a dot in every loop of the network whose dual is obtained and a dot outside the network. Each dot is numbered according to the loop in which it is placed. The outside dot is called the reference node and give number as 0.
- Connect two dots by a line through each branch. The dots are the nodes of the dual network between two nodes; the element to be connected is the dual of the element crossed by the line.
- When sources are included, then the line joining the dots should intersect the sources also; between these two nodes the dual of the source is included.
- The polarity of the source is decided by the fallowing rule. A voltage or current source which drives a current in clockwise in i^{th} loop, then place a positive polarity at the dual network. Negative if it is opposite.

Problems

1. Let us draw the dual network of the circuit shown in the below figure.

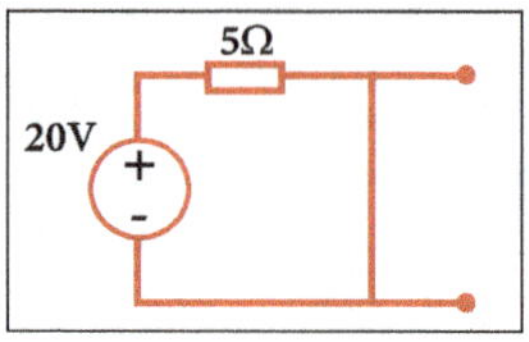

Solution:

Step 1:

The dual of 20V voltage source is 20A current source as shown in the below figure.

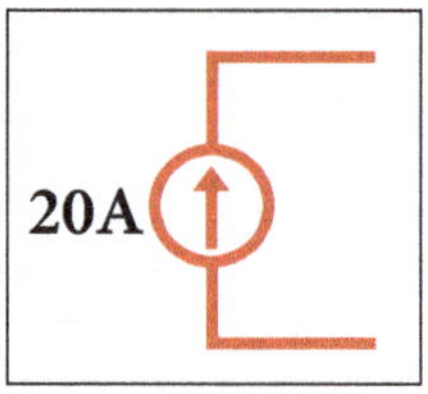

Step 2:

The dual of 5Ω resistor is 1/5 series = 1/5Ω resistor in parallel.

Step 3:

Since the 20V voltage source is in series with 5Ω resistor, its dual counterpart is 20A current source which is in parallel with 1/5Ω. This is shown in the below figure.

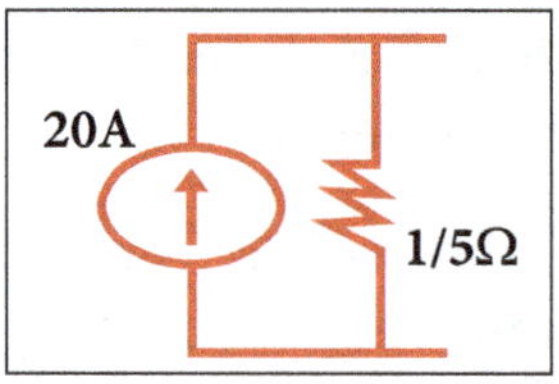

2. Let us draw the dual network of below figure.

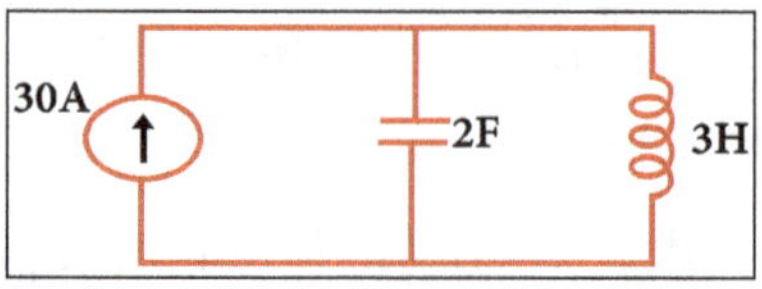

Solution:

Since the current source, capacitor and inductor are all connected in parallel, their dual counterparts are connected in series.

The dual of 30A current source is 30V voltage source.

The dual of 2F capacitor is 2H inductor.

The dual of 3H inductor is 3F capacitor.

The dual network is obtained as follows.

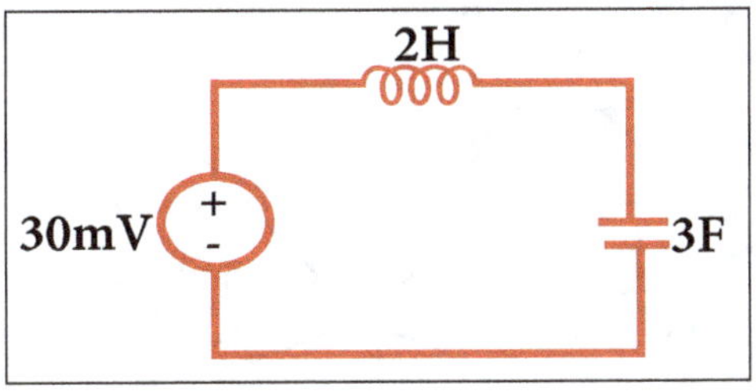

3. Let us draw the dual network for the given network.

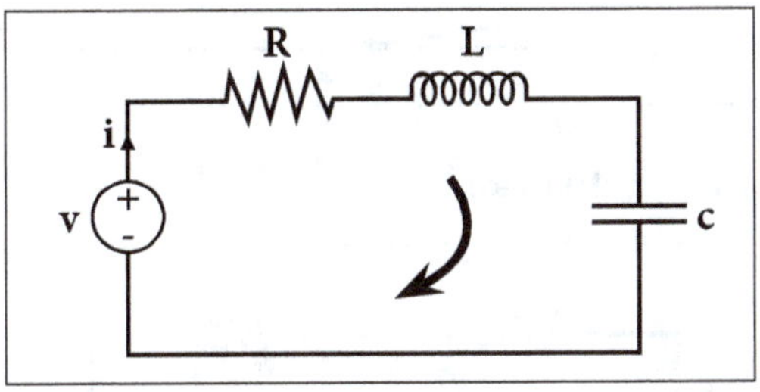

Solution:

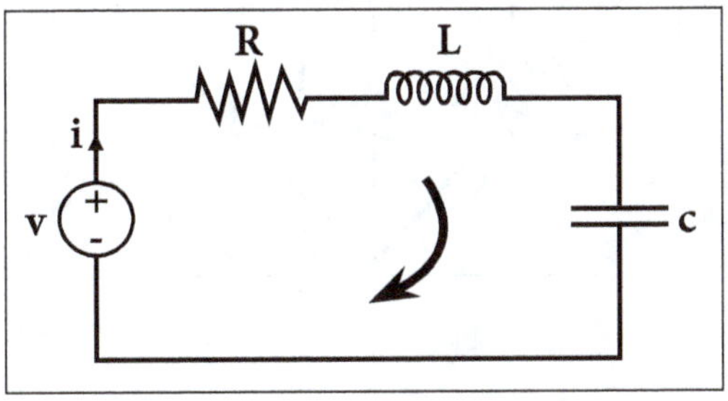

The integral differential equations for the circuit are given as:

$$R\,i + L\,(di/dt) + (1/C)\int i\,dt = v$$

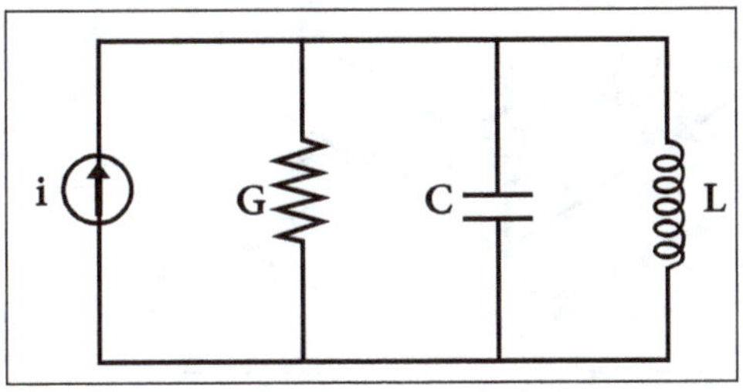

Dual Network:

The integral differential equations for the circuit is given as:

$$(1/R)\,v + C\,(dv/dt) + (1/L)\int v\,dt = i$$

or

$$G\,v + C\,(dv/dt) + (1/L)\int v\,dt = i$$

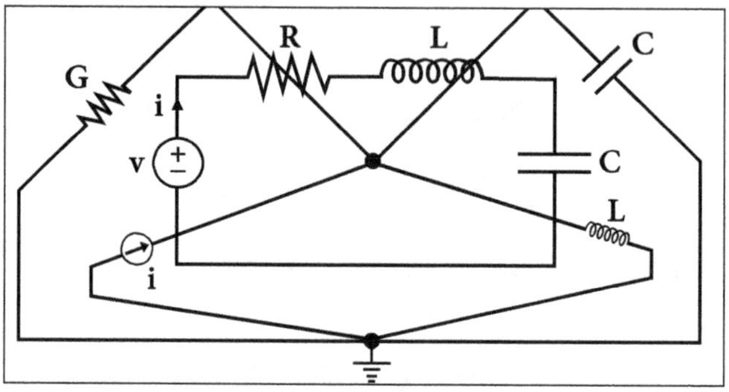

4. Let us draw the dual of the network shown in the below figure.

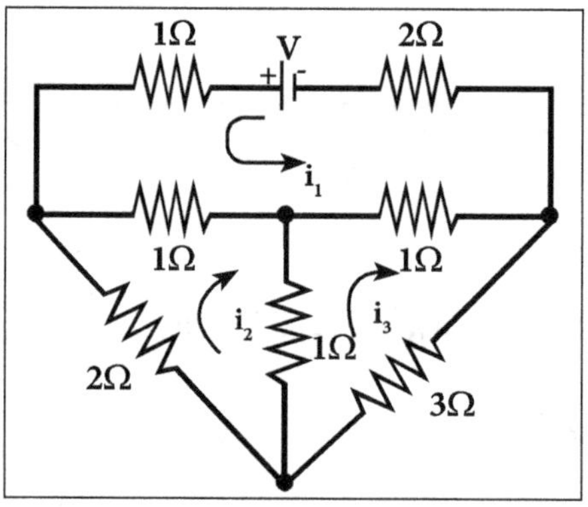

Solution:

The dual network of above figure is shown in the below figure.

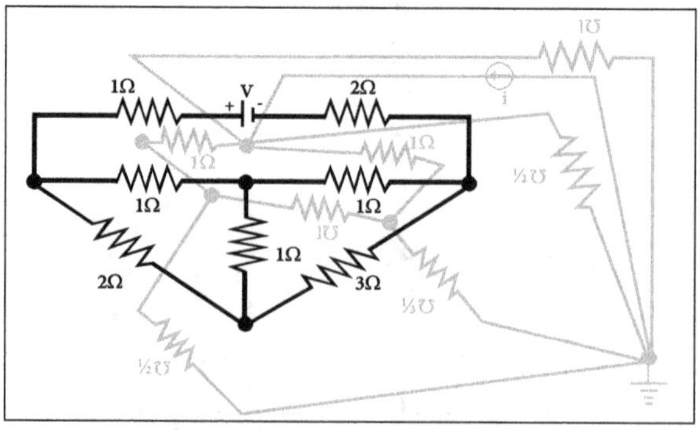

5. For the network shown in the below figure, let us write the node equations. Let us draw the dual of this network and write mesh equations for the dual network and let us also verify whether these two sets of equations are dual equations.

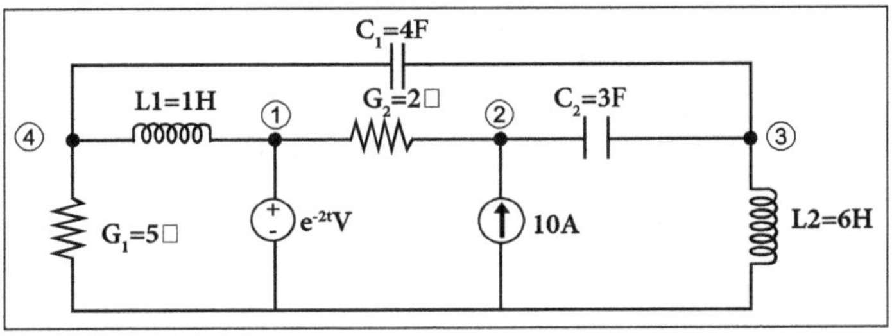

Solution:

Node equation:

Node (1),

$$(1/L_1)\int\left(e^{-2t}-v_4\right)dt+G_2\left(e^{-2t}-v_2\right)=0 \quad ...(1)$$

Node (2),

$$(G_2)\int\left(v_2-e^{-2t}\right)+C_2\left[d\left(v_2-v_3\right)/dt\right]+10=0 \quad ...(2)$$

Node (3), and

$$(1/L_2)\int v_3\,dt+C_2\left[d\left(v_3-v_2\right)/dt\right]+C_1\left[d\left(v_3-v_4\right)/dt\right]=0 \quad ...(3)$$

Node (4),

$$G_1v_4+(1/L_1)\int\left(v_4-e^{-2t}\right)dt+C_1\left[d\left(v_4-v_3\right)/dt\right]=0 \quad ...(4)$$

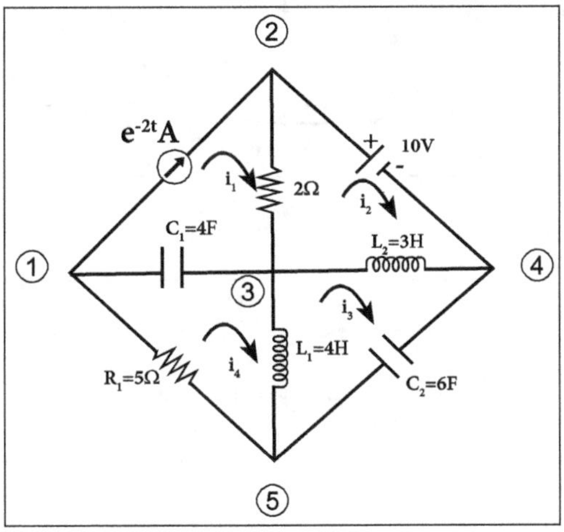

Loop equation for the dual network is given as:

$$(1/C_1)\int\left(e^{-2t}-i_4\right)dt+R_2\left(e^{-2t}-i_2\right)=0 \quad ...(5)$$

$$R_2\int\left(i_2-e^{-2t}\right)+L_2\left[d\left(i_2-i_3\right)/dt\right]+10=0 \quad ...(6)$$

$$(1/C_2)\int i_3\,dt+L_2\left[d\left(i_3-i_2\right)/dt\right]+L_1\left[d\left(i_3-i_4\right)/dt=0\right] \quad ...(7)$$

$$R_1 i_4+(1/C_1)\int\left(i_4-e^{-2t}\right)dt+L_1\left[d\left(i_4-i_3\right)/dt\right]=0 \quad ...(8)$$

By looking into two sets of equations, we can say that two sets of equations are dual equations.

6. Let us obtain the dual of the network shown in the below figure.

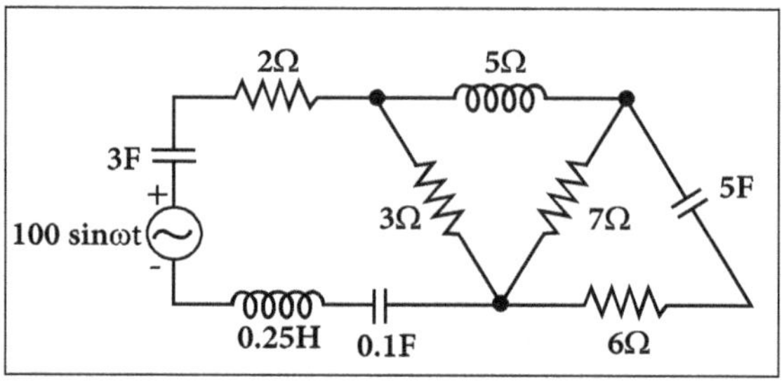

Solution:

Putting nodes inside identified independent loops and considering one datum node outside the network, the elements are traced (mapped) as follows:

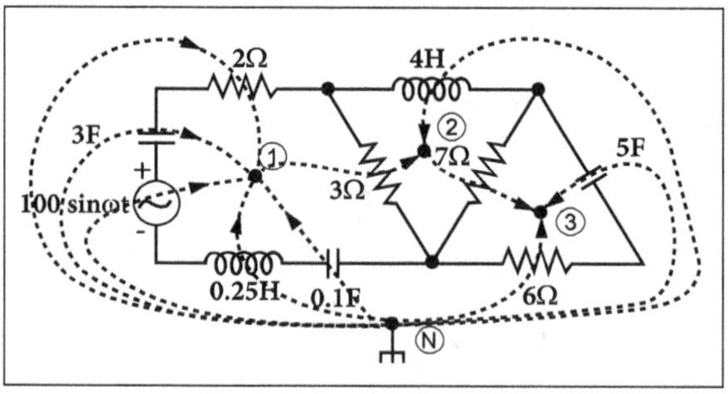

The dual network of the above network is as follows:

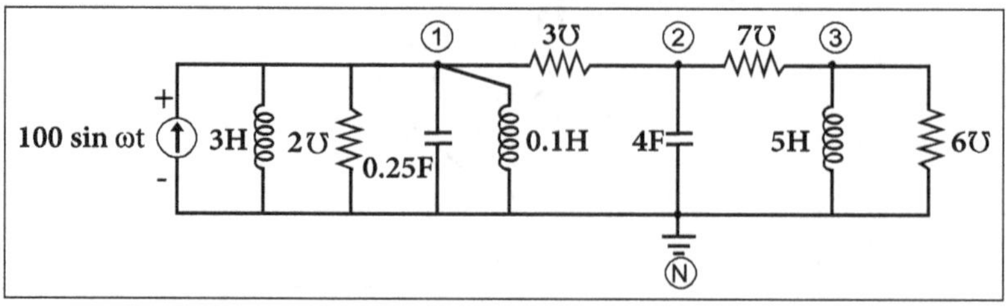

2

Network Theorems for DC and AC Circuits

2.1 Network Theorems

Network

It is a collection of interconnected components. Network analysis is the process of finding the voltages across and the current through every component in the network. There are many different techniques for calculating these values.

The most important fundamental theorems of network analysis are as follows:

- Thevenin's theorem
- Norton's theorem
- Millman's theorem
- Reciprocity theorem
- Substitution theorem
- Superposition theorem
- Maximum power transfer theorem

2.1.1 Superposition Theorem

It is one of the electrical network analysis theorem which helps us to solve the linear circuit with more than one current or voltage source.

In a linear circuit with several sources, the current and voltage responses in any branch is the algebric sum of the voltage and current responses due to each of the source acting independently where all the other sources are replaced by their internal impedance.

Let us consider an electrical circuit having several branches or loads and also several source namely current source and voltage source.

If we find the branch responses on a branch due to only of those source by ignoring the effect of all the other sources or replacing all the other sources by their corresponding internal impedance and the process for every source on the circuit is repeated, then the combined responses on a branch due to all the sources combined is the algebric sum of the responses on the branches due to each individual sources.

Steps to be followed to apply the superposition theorem:

- Step 1: Any one of the energy source is selected and all the other energy sources are replaced by their internal series resistances.
- Step 2: With only one energy source, let us calculate the current through the required branch with current division rule or using mesh analysis.
- Step 3: Steps 1 and 2 is repeated for each source individually.
- Step 4: Currents obtained due to the individual source are added to obtain the combined effect of all the sources.

2.1.2 Thevenin's Theorem

In circuit theory, Thevenin's theorem for linear electrical networks states that, any combination of voltage sources, current sources and resistors with two terminals is electrically equivalent to the single voltage source V and the single series resistor R. For a single frequency AC systems, the theorem can also be applied to general impedances.

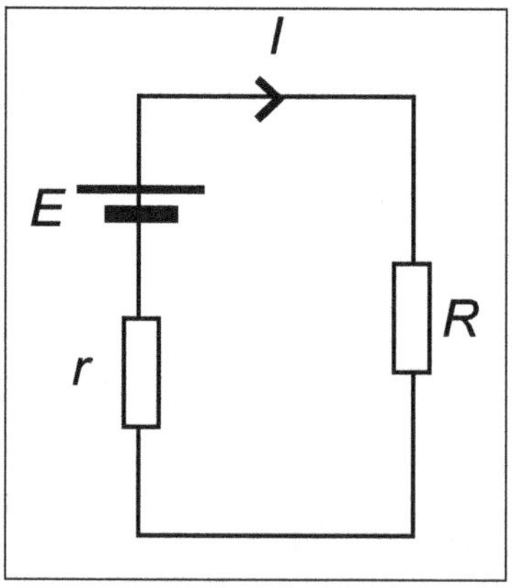

Thevenin's circuit.

The procedure adopted when using the Thevenin's theorem is summarized as follows.

To determine the current in any branch of the active network, following procedures are followed:

- The resistance R from the branch is removed.
- Open-circuit voltage (E) across the break is determined.

- Each of the sources of e.m.f. is removed and they are replaced by their internal resistances to determine the resistance (r).
- Current from the equivalent circuit is determined as follows.

$I = E_R + r$

Problems

1. Let us determine the current I_{ab} flowing through the 3Ω resistor as shown in the below figure using the superposition theorem.

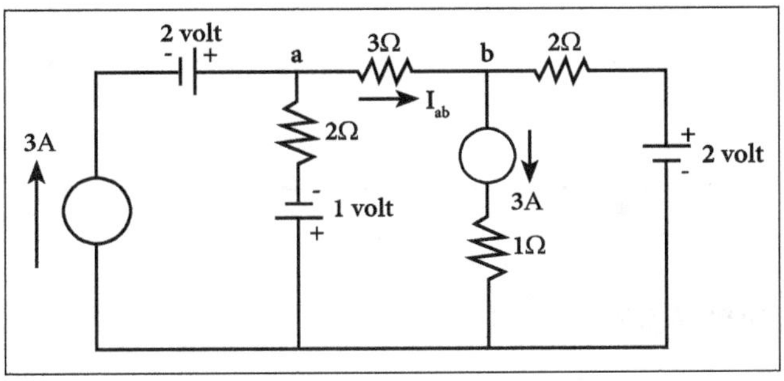

Solution:

Given:

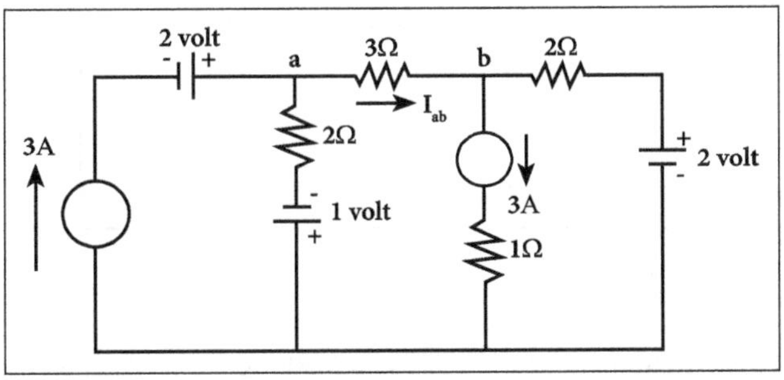

(a)

Let us assume that the current source 3A is acting alone in the circuit and the internal resistances are replaced by the other sources.

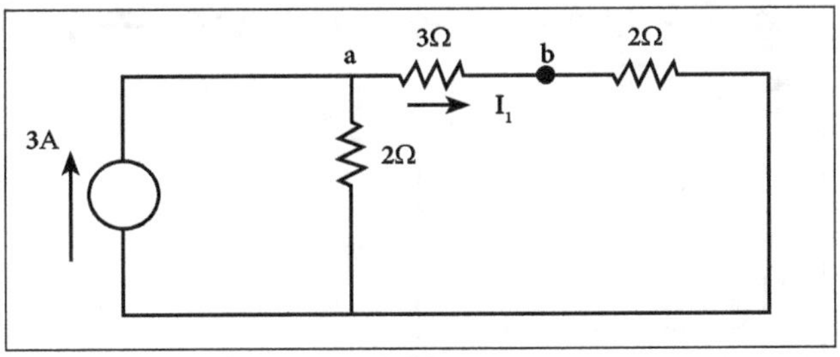

(b)

The current flowing through 3 Ω resistor is obtained from figure (b) as follows.

$$I_{1(\text{due to 3A current source})} = 3 \times \frac{2}{7} = \frac{6}{7}\ \text{A}(\text{a to b}) \quad ...(1)$$

Current flowing through 3Ω resistor due to 2V source is obtained from the below figure(c) as follows.

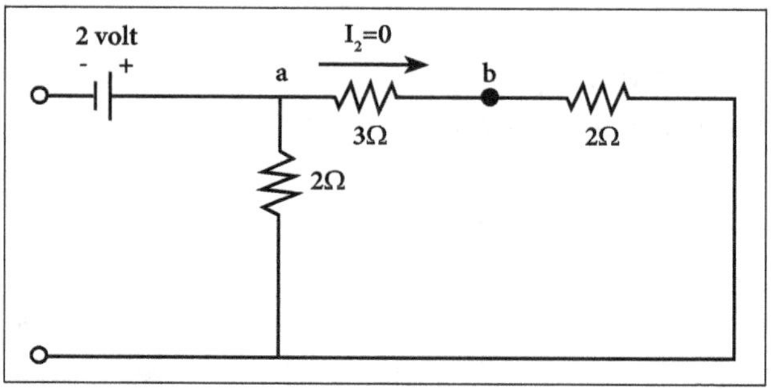

(c)

Since current is not flowing, we have:

$$I_{2(\text{due to 2V voltage source})} = 0\text{A}\ (\text{a to b})$$

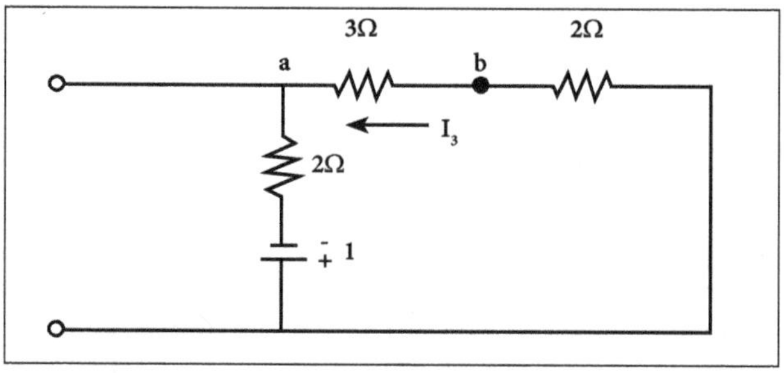

(d)

Current through 3Ω resistor due to 1V voltage source is given as:

$$I_{3(\text{due to 1V voltage source})} = 1/7\ \text{A}(\text{b to a})$$

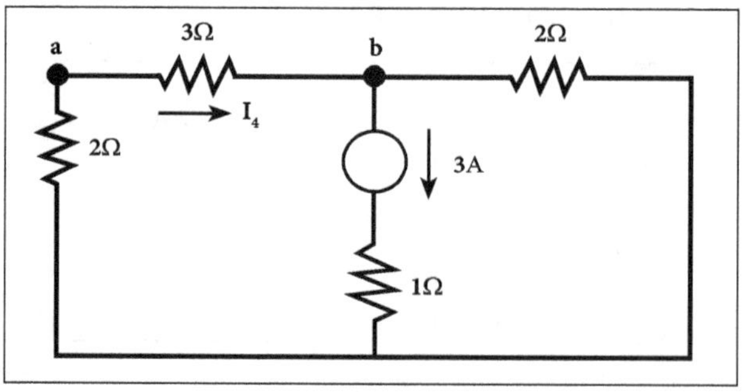

(e)

From the above figure (e), current through 3Ω resistor due to 3A current source is given as:

$$I_{4(\text{due to 3A current source})} = 3 \times \frac{2}{7} = \frac{6}{7}\ A(\text{a to b})$$

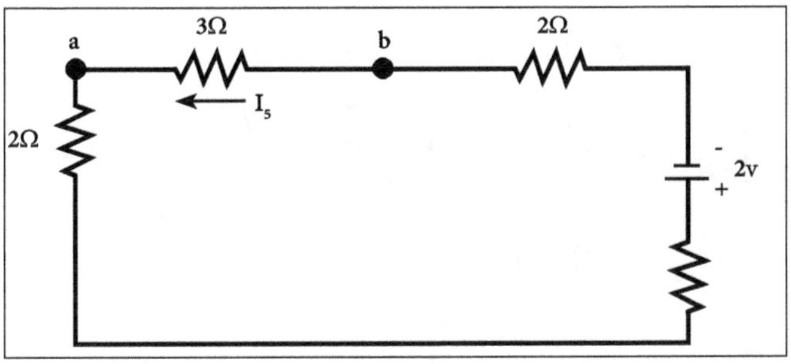

(f)

From the above figure (f), current through 3 Ω resistors due to 2V voltage source only is given as:

$$I_{5(\text{due to 2V voltage source})} = \frac{2}{7}\ A(\text{b to a}) \qquad \text{...(2)}$$

Resultant current I_{ab} flowing through 3Ω resistor due to the combination of all sources is obtained by the following expression.

$$I_{ab} = I_{1(\text{due to 3A current source})} + I_{2(\text{due to 2V current source})} + I_{3(\text{due to 1V current source})}$$
$$+ I_{4(\text{due to 3A current source})} + I_{5(\text{due to 3A current source})}$$

$$= \frac{6}{7} + 0 - \frac{1}{7} + \frac{6}{7} - \frac{2}{7} = \frac{9}{7} = 1.285\ (\text{a to b})$$

2. Using the superposition theorem, let us determine the voltage drop and current across the resistor 3.3KΩ as shown in the below figure.

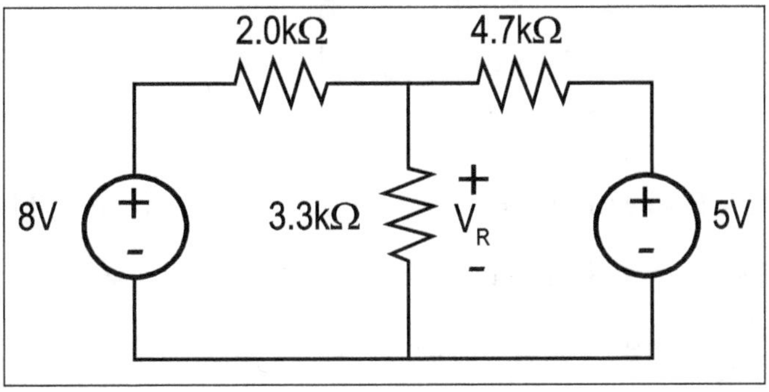

Solution:

Step 1:

Remove the 8V power supply from the original circuit where the new circuit is shown in the below figure and then the voltage across a resistor is measured.

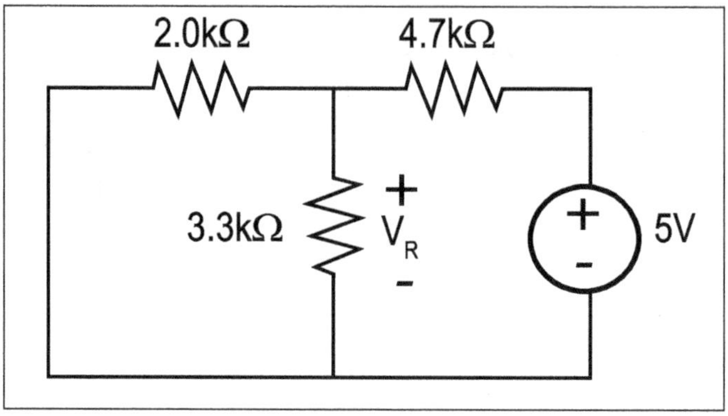

Since 3.3KΩ and 2KΩ are in parallel, the resultant resistance will be 1.245KΩ.

Using voltage divider rule, voltage across 1.245KΩ is given as:

$$V_1 = \left[1.245/(1.245+4.7)\right]*5 = 1.047V$$

Step 2:

5V power supply is removed from the original circuit such that the new circuit is shown in the below figure and the voltage across a resistor is measured.

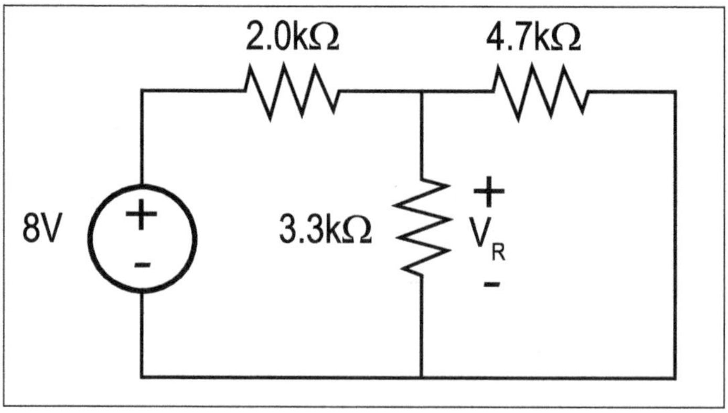

Since 3.3KΩ and 4.7KΩ are in parallel, the resultant resistance will be 1.938KΩ.

Using voltage divider rule, voltage across 1.938KΩ is given as:

$$V_2 = \left[1.938/(1.938+2)\right]*8 = 3.9377V$$

Voltage drop across a 3.3KΩ resistor is given as:

$$V_1 + V_2 = 1.047 + 3.9377 = 4.9847V$$

3. Let us find the current through 10 Ω resistance in the given network by the using superposition theorem.

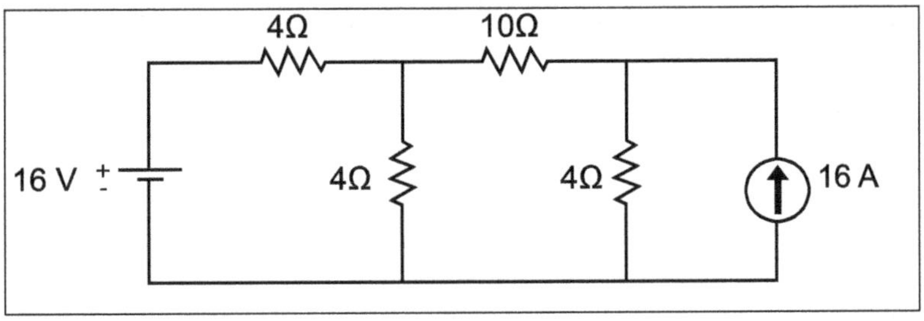

Solution:

Step 1:

Current through 10Ω resistance is found by using superposition theorem.

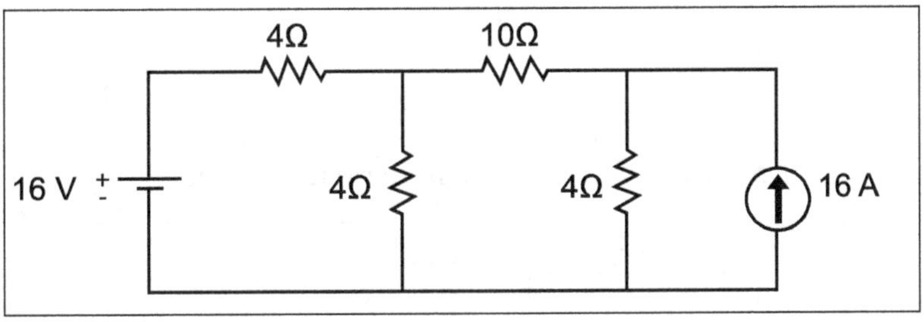

If '16V' source is activated, then the other will be deactivated.

Step 2:

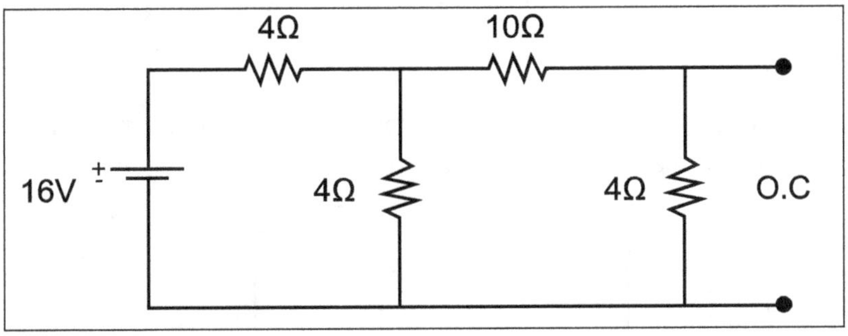

After deactivation of '16A' current source, two individual boxes are considered from which we can easily apply mesh analysis for finding current through 10Ω resistance where '16V' voltage source is active. We can use nodal analysis or ohm's law with current division rule.

Step 3:

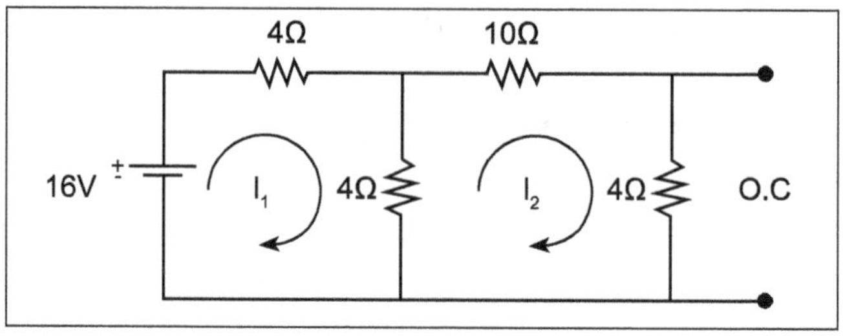

Mesh equations are given as:

$$-8I_1 + 4I_2 = -16 \quad ...(1)$$

$$4I_1 - 18I_2 = 0 \quad ...(2)$$

After solving this circuit, we have:

$$I_1 = 2.25A \ \& \ I_2 = 0.5A$$

$$I_{10\Omega} = I_2 = 0.5 \text{ A}$$

Step 4:

If '16V' source is activated, then the other will be deactivated.

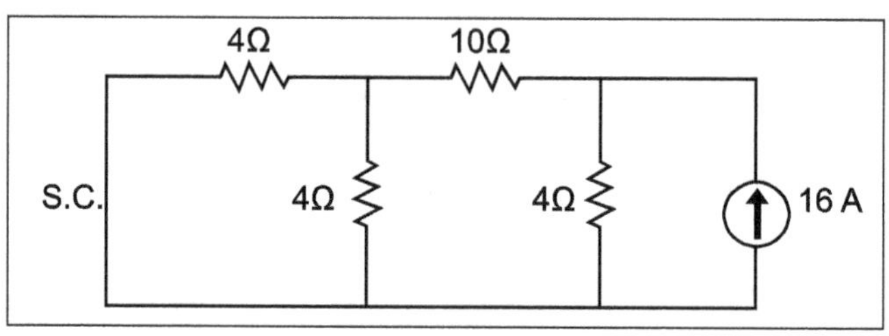

Step 5:

After deactivation of '16V' voltage source, current division rule is applied. By applying current division rule, we can easily find the value of current in 10Ω resistance.

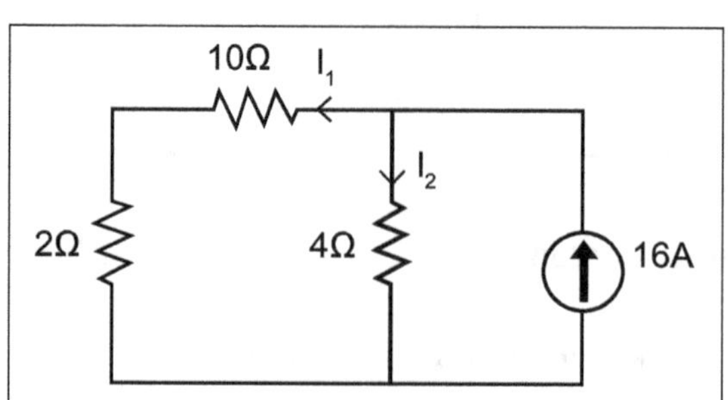

By applying current division rule, we have:

$$I_1 = 4A$$

$$I_2 = 12A$$

$$I_{10\Omega} = I_1 = 4A$$

Step 6:

In the final circuit, all the sources are activated. Direction of current when single source is activated and other is deactivated through 10Ω resistance is shown in the below figure.

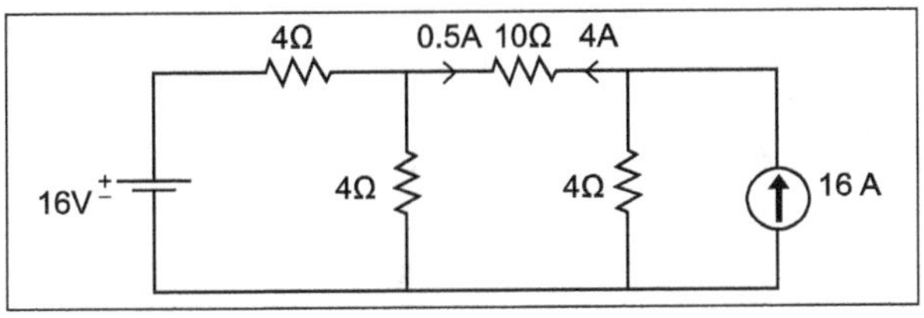

Step 7:

The current through 10Ω resistance is 3.5A.

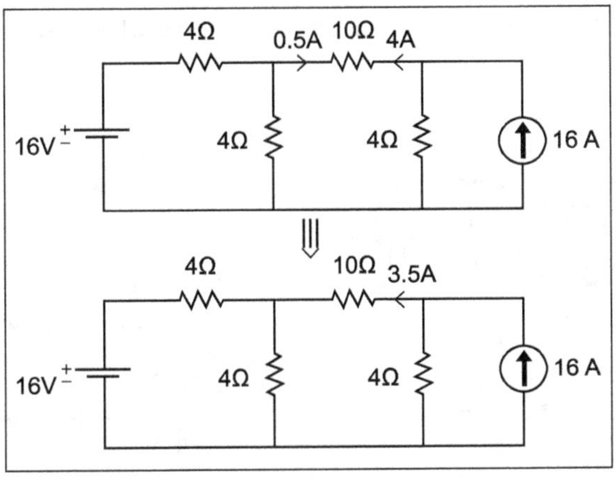

Current flowing through 10 Ω = 3.5 A.

4. Let us determine the current in the 0.8Ω resistor using Thevenin's theorem for the below circuit.

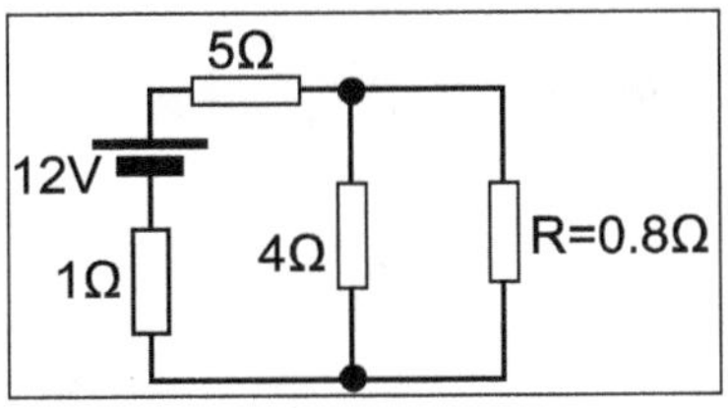

Solution:

Given:

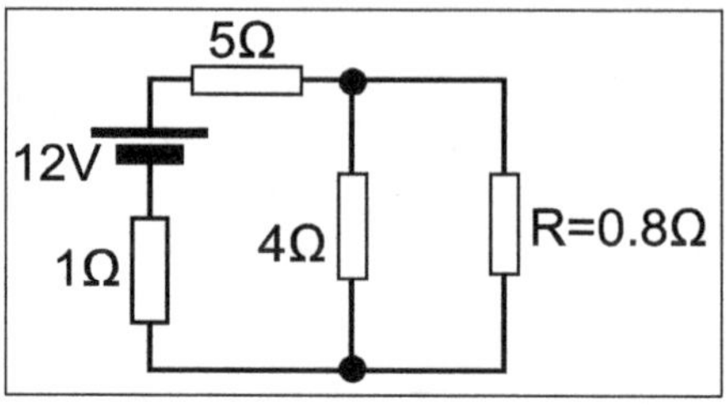

(i) 0.8Ω resistor is removed from the circuit as shown in the above figure.

Current is given as:

$$I_1 = 12/(1+5+4)$$

$$= 12/10$$

$$= 1.2A$$

(ii) Power dissipation across 4Ω resistor = $4I_1$ = (4) (1.2) = 4.8V.

Hence, power dissipation across AB i.e., the open-circuit voltage across AB is given as:

E = 4.8 V

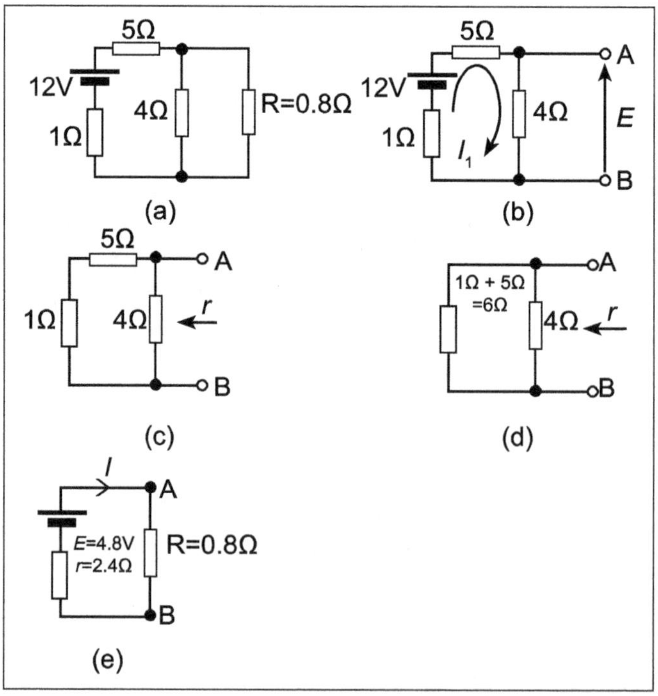

(iii) Circuit shown in the above figure (c) is obtained by removing the source of e.m.f. The equivalent circuit of figure(c) is shown in figure (d) where:

$$\text{Resistance } (r) = (4 \times 6)/(4 + 6) = 24/10 = 2.4\,\Omega$$

(iv) Equivalent Thevenin's circuit is shown in the above figure (e).

From the above circuit (e), we have:

$$\text{Current } I = E/(r + R)$$

$$= 4.8/(2.4 + 0.8)$$

$$= 4.8/3.2 \text{ A}$$

$$I = 1.5\text{A} = \text{Current inthe } 0.8\,\Omega \text{ resistor}$$

5. Let us find the Thevenin's equivalent at terminals a-b of the below circuit.

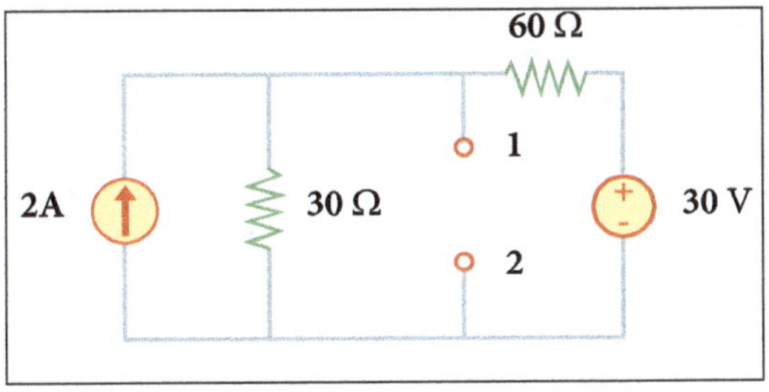

Solution:

Given:

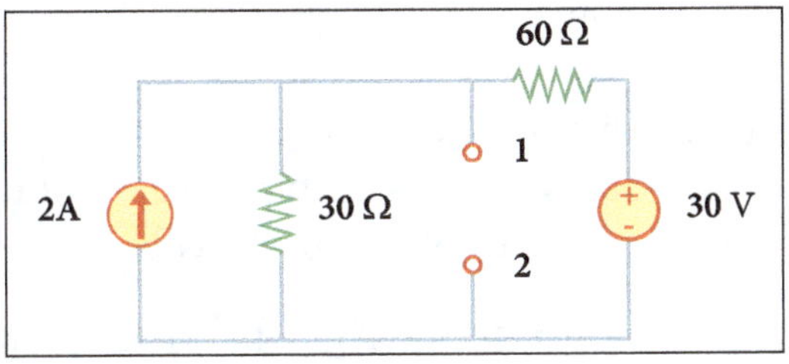

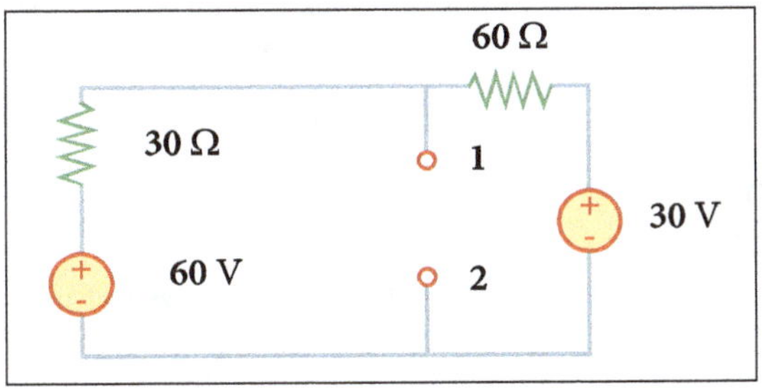

If we perform the source transformation, the original circuit changes to a simple series one where two sources are in the opposite direction.

$$V_{sigma} = 60V - 30V = 30V$$

$$R_{sigma} = 30\Omega + 60\Omega = 90\Omega$$

$$I_{sigma} = V_{simga} / R_{sigma} = 1/3\,A$$

$$V_{12} = V_{60} + 30V = 1/3 * 60 + 30V = 50V$$

6. Using Thevenin's theorem calculate the range of current flowing through the resistance R when its value is varied from 6 Ω to 36 Ω.

Solution:

The open circuit voltage, V_{OC} across the terminals A and B by removing the resistance R is calculated as:

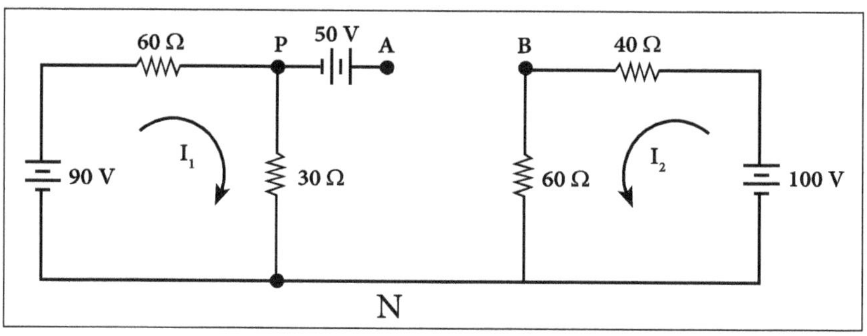

$$I_1 = \frac{90}{60+30} = 1A,\ I_2 = \frac{100}{60+40} = 1A$$

Voltage drop across the 30Ω resistor = 30 x 1 = 30 V. The potential of point P with respect to N is + 30 V. The potential of point A with respect of point P is +50 volts. Therefore, the to potential of point A with respect to point N is +30 + 50 = +80 V.

Voltage drop across the 60 Ω resistor in the right-hand-side loop is calculated as:

$$I_2 = \frac{100}{40+60} = 1A$$

$$V_{NB} = 60 \times 1 = 60\,V$$

Potential of point B with respect to point N is + 60 V and that of A w.r.t. N = $30I_1 + 50$ = +80V.

Now we observe that the potential of point A with respect to point N is +80 V and the potential of point B with respect to point N is +60 V. Therefore, the potential of point A with respect to point B, i.e., V_{OC} becomes +20 V, point A being at higher a potential than point B.

To calculate R_{eq} we redraw the circuit by short circuiting the sources of EMFs as in the below figure.

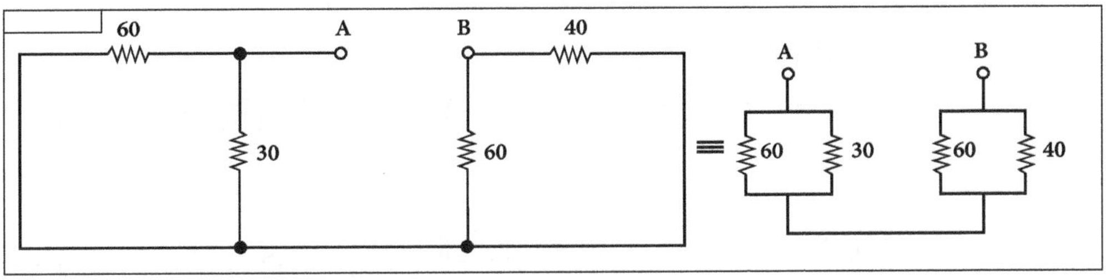

$$R_{eq} = \frac{60\times 30}{60+30} + \frac{60\times 40}{60+40} = 44\ \Omega$$

Thevenin's equivalent circuit is drawn as in the below figure.

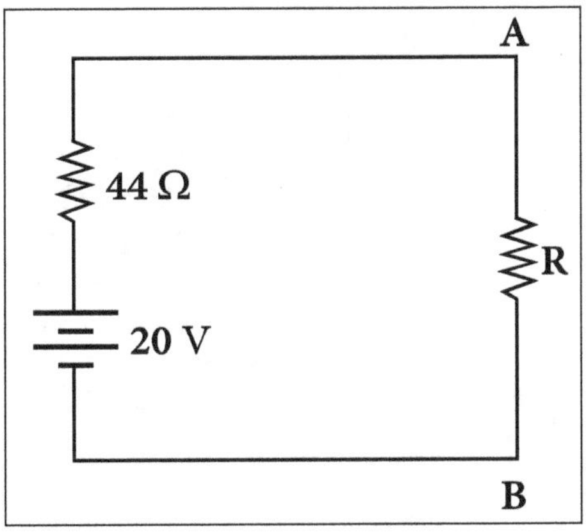

$$I = \frac{20}{44+6} \text{ when } R = 6\Omega$$

$$= 0.4\text{A}$$

$$I = \frac{20}{44+36} \text{ when } R = 36$$

$$= 0.25\text{ A}$$

The value of current through the resistor R will vary from 0.4 A to 0.25 A when its value is changed from 6 Ω to 36 Ω keeping the other circuit conditions unchanged.

7. Using Thevenin's theorem, let us calculate the current flowing through the load resistance R_L connected across the terminals A and B as shown in the below figure.

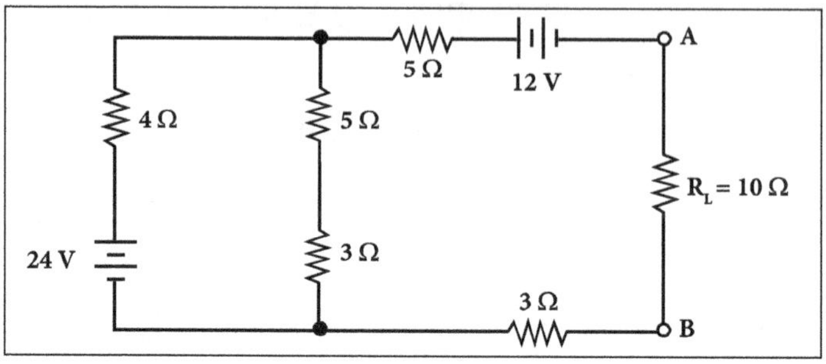

Solution:

First we remove the load resistance R_L from the circuit and calculate the open circuit voltage, V_{OC} across its terminals A and B as:

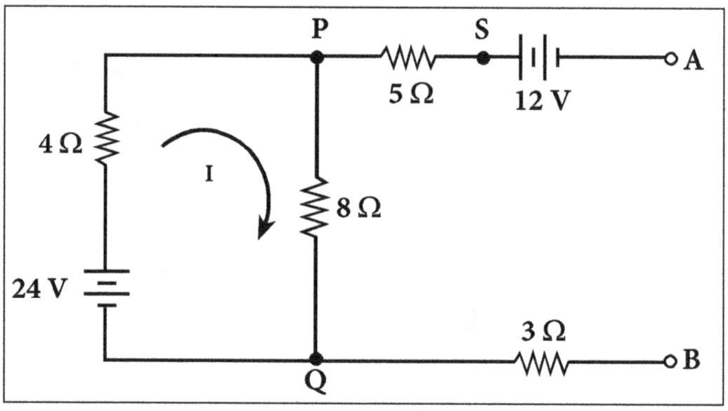

Let current I be known through the loop as shown.

$$I = \frac{24}{4+8} = 2\,A$$

Voltage drop across PQ is equal to I_R = 2 x 8 = 16 V. Terminal P is at higher potential than terminal Q. No current is flowing through the 3 Ω and 5 Ω resistors in the circuit because these are open circuited.

The potential of A with respect to B is calculated starting from point B as:

$$V_{AB} = V_{BQ} + V_{QP} + V_{PS} - V_{SA}$$

$$= 0 + 16\,V + 0 - 12\,V$$

$$= +4\,V$$

$$= V_{OC}$$

For calculating R_{eq}, we short circuit the sources of EMFs as:

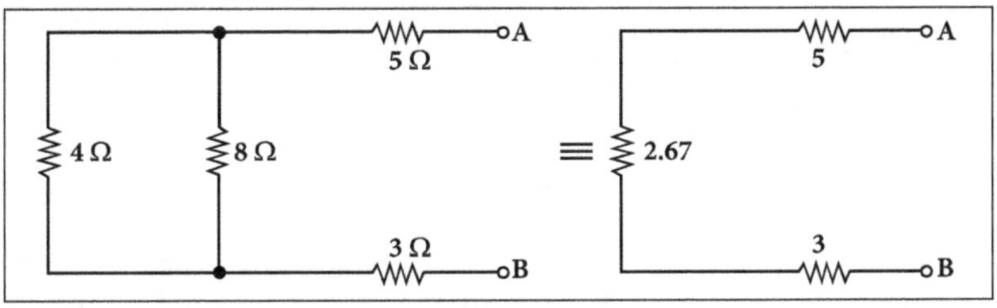

$$R_{eq} = R_{AB} = 10.67\ \Omega$$

Thevinin's equivalent circuit and current through the load resistor is calculated as:

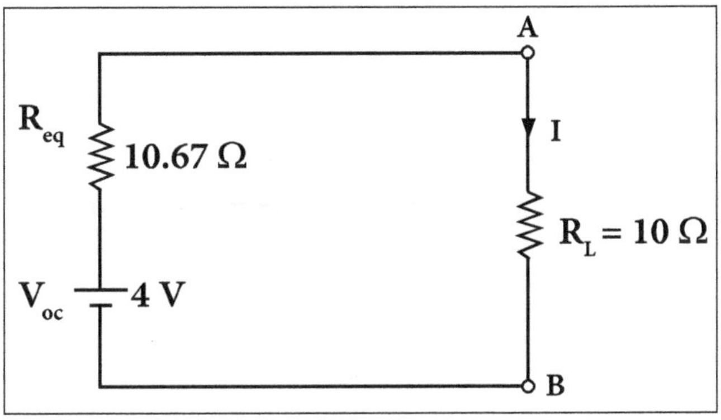

$$I = \frac{V_{OC}}{R_{eq} + R_L}$$

$$= \frac{4}{10.67 + 10} = 0.193\ A$$

2.1.3 Norton's Theorem

Any two-terminal linear bilateral DC network can be replaced by an equivalent circuit consisting of a current and a parallel resistor.

Preliminary Steps

The steps leading to the proper values of I_N and R_N are as follows:

- Portion of the network is removed across which the Norton equivalent circuit is found.
- Terminals of the remaining two-terminal network are marked.
- R_N is found as follows.

Let us calculate the value of R_N by setting all sources to zero and then the resultant resistance between the two marked terminals are found. Since $R_N = R_{Th}$, the procedure and value obtained using the approach described for Thevenin's theorem will determine the proper value of R_N.

- I_N is found as follows.

  Let us calculate the value of I_N by returning all the sources to their original position and then the short-circuit current between the marked terminals is found. It is the same amount of current that can be measured by an ammeter which is placed between the marked terminals.

- Finally, Norton equivalent circuit is drawn by replacing the terminals of the equivalent circuit.

Norton's Equivalent Circuit

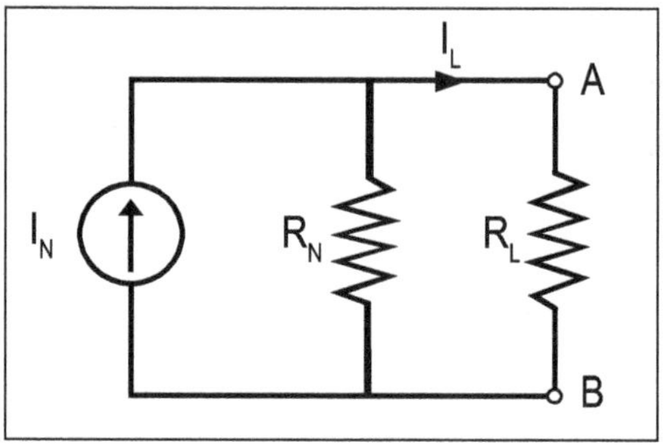

The current through R_L is given as:

$$I_L = \frac{I_N \cdot R_N}{R_N + R_L}$$

2.1.4 Reciprocity Theorem

It is one of the theorem used for DC network analysis and AC network analysis technique which deals with the relationship between impressed source in a part of the circuit and its response at some other part of the circuit.

The supply voltage "V" and a 0-ohm ammeter reading or current "I" are mutually transferable in any bilateral linear circuits. The ratio between the voltage and the current which are mutually transferable is called as the transfer resistance.

It is true in bilateral linear circuits or on the linear circuits which contains only bilateral components. This property of bilateral linear circuits is termed as their reciprocal property and the circuits in which the reciprocity theorem holds true are termed as reciprocal circuits.

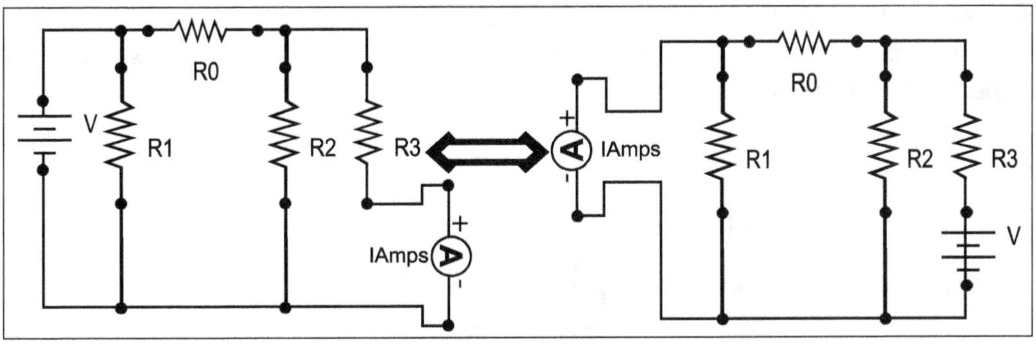

Reciprocity theorem.

It is applied to solve many DC and AC electrical network which have special applications in the field of electromagnetism and antenna electronics.

2.1.5 Millman's Theorem

It is possible to combine the number of voltage sources or current sources into a single equivalent voltage or current source using Millman's theorem. Therefore, this theorem is used for calculating the total current supplied to the load in a generating station by a number of generators that are connected in parallel across a bus bar.

Millman's theorem states that, if n number of generators having generated emf's E_1, E_2, ..., E_n and internal impedances Z_1, Z_2, ..., Z_n are connected in parallel, then the emfs and impedances are combined to give a single equivalent emf(E) with an internal impedance(Z).

$$E = \frac{E_1 Y_1 + E_2 Y_2 + \ldots + E_n Y_n}{Y_1 + Y_2 + \ldots + Y_n}$$

$$Z = \frac{1}{Y_1 + Y_2 + \ldots + Y_n}$$

Where,

Y_1, Y_2, ... Y_n = Admittances corresponding to the internal impedances Z_1, Z_2, ..., Z_n

$$Y_1 = \frac{1}{Z_1}$$

$$Y_2 = \frac{1}{Z_2}$$

$$\vdots$$

$$Y_n = \frac{1}{Z_n}$$

The below figure shows a number of generators having emfs E_1, E_2, ..., En which are connected in parallel across the terminals x and y. Z_1, Z_2, ..., Z_n are the respective internal impedances of the generators.

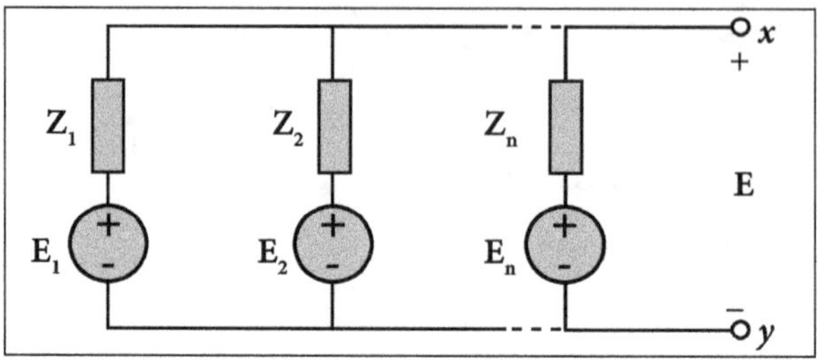

Thevenin's equivalent circuit of the above figure using Millman's theorem is shown in the below figure.

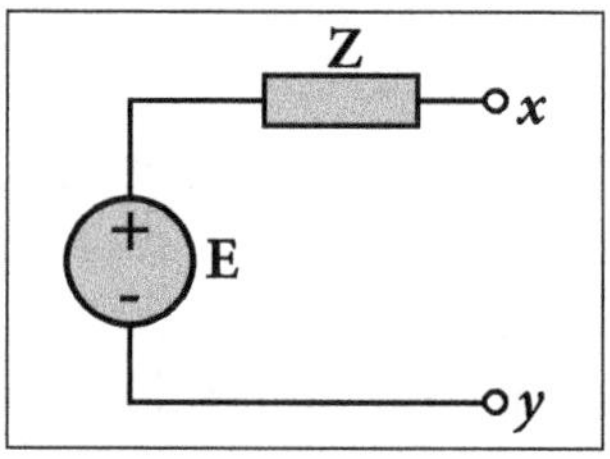

Thevenin's equivalent circuit.

The nodal equation at x is given as:

$$\frac{E_1 - E}{Z_1} + \frac{E_2 - E}{Z_2} + ... + \frac{E_n - E}{Z_n} = 0$$

$$\Rightarrow \left[\frac{E_1}{Z_1} + \frac{E_2}{Z_2} + ... + \frac{E_n}{Z_n}\right] = E\left[\frac{1}{Z_1} + \frac{1}{Z_2} + ... + \frac{1}{Z_n}\right]$$

$$\Rightarrow E_1 Y_1 + E_2 Y_2 + ... + E_n Y_n \left[\frac{1}{Z}\right]$$

Where,

Z = Equivalent internal impedance

$$\text{or, } [E_1 Y_1 + E_2 Y_2 + ... + E_n Y_n] = EY$$

$$\Rightarrow E = \frac{E_1 Y_1 + E_2 Y_2 + ... + E_n Y_n}{Y}$$

$$\text{where, } Y = Y_1 + Y_2 + ... + Y_n$$

$$\text{and,}\ Z = \frac{1}{Y} = \frac{1}{Y_1 + Y_2 + ... + Y_n}$$

Problems

1. Let us determine the current I flowing in the 4Ω resistance shown in the below figure by using Norton's theorem.

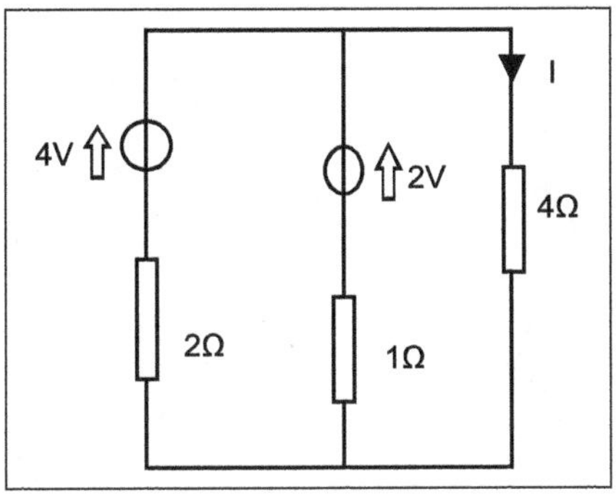

Solution:

Given:

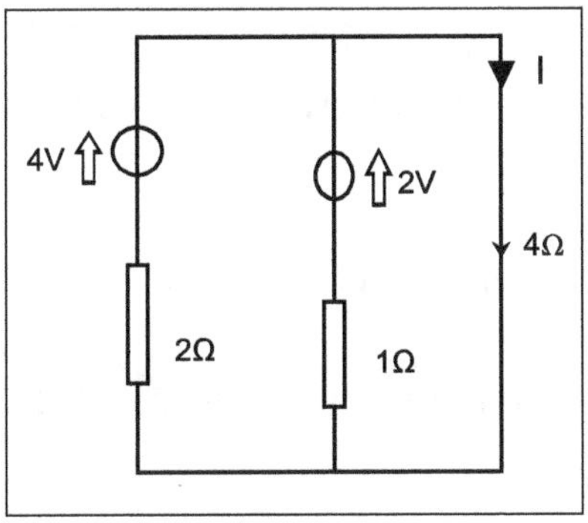

4Ω branch is short-circuited as shown in the below figure (b).

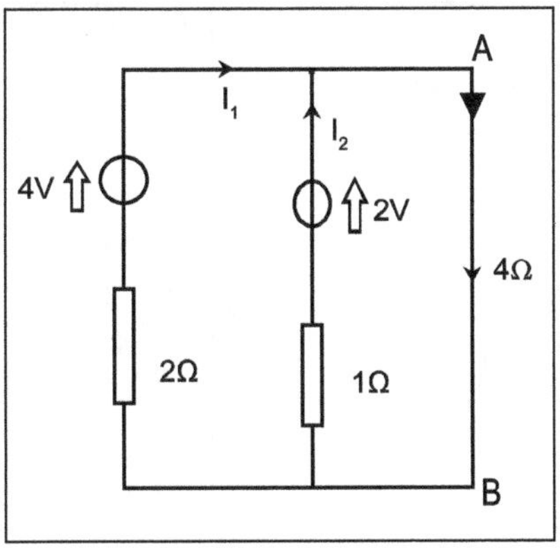

From the above figure, we have:

$$I_{SC} = I_1 + I_2 = 4A$$

If the sources of e.m.f. are removed, then the resistance between A and B is given as:

$$r = 2\times1/2+1 = 2/3\,\Omega$$

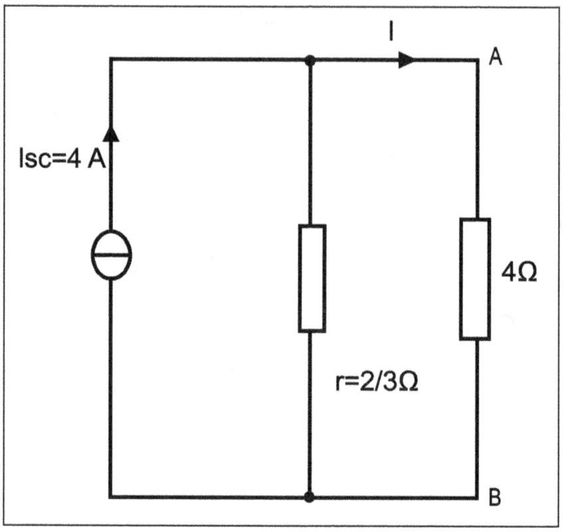

From the Norton equivalent network shown in above figure, the current in the 4 Ω resistance is given as:

$$I = \left[(2/3)/((2/3)+4)\right](4) = 0.571\,A$$

2. Let us find the Norton's equivalent circuit across A-B terminals for the below circuit.

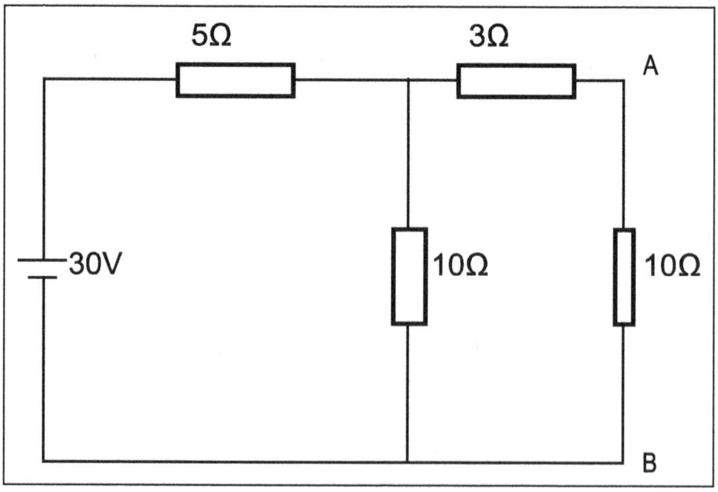

Solution:

10Ω resistor is removed and the terminals A & B are short circuited as shown in the below figure.

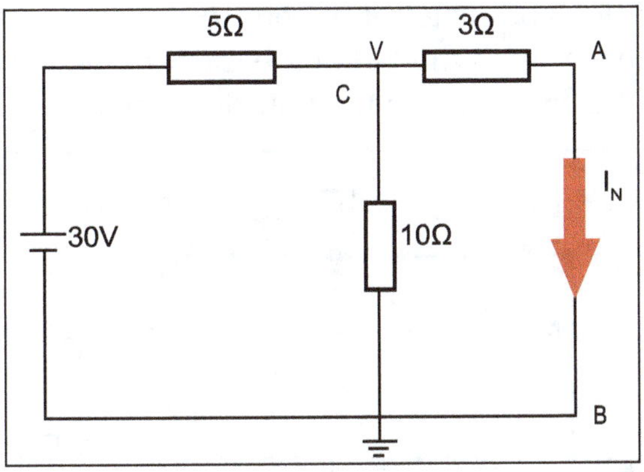

The current flowing through the short circuited terminals is called as the Norton's current (I_N).

To find the I_N, let us apply the nodal equation at point C.

$$(V-30)/5+V/10+V/3=0$$

$$V = 180/19V$$

Applying ohm's law to 3Ω resistance, we have:

$$I=V/R$$

$$I_N=(180/19)/3A$$

$$I_N=60/19A$$

This is known as the Norton's current.

Let us find the Norton's resistance which is equal to the thevenin's resistance.

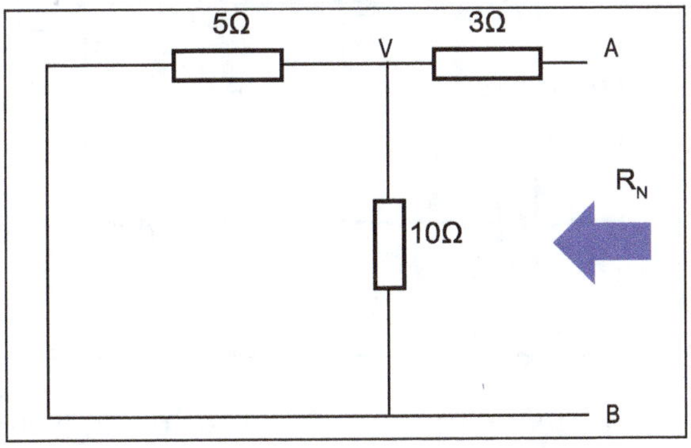

$$R_N = (10/5)\,\Omega + 3\Omega$$

$$R_N = 19/3\,\Omega$$

3. The current in the various branches of the network is shown in the below figure. Let us determine the current through the one volt battery (V) when an extra e.m.f of 1 volt is added on the branch BD by opposing the flow of original current in that branch.

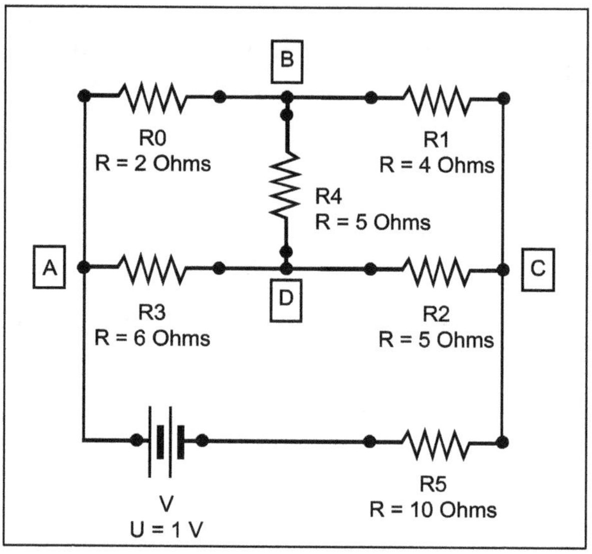

Solution:

Given:

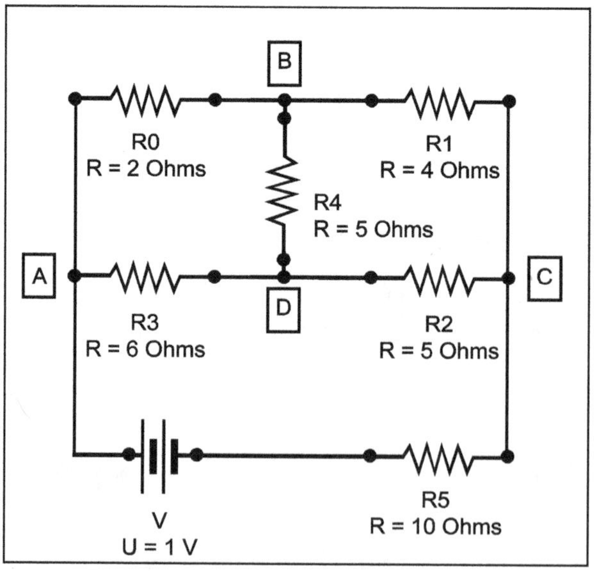

The above circuit is solved by using the mesh analysis or loop current method.

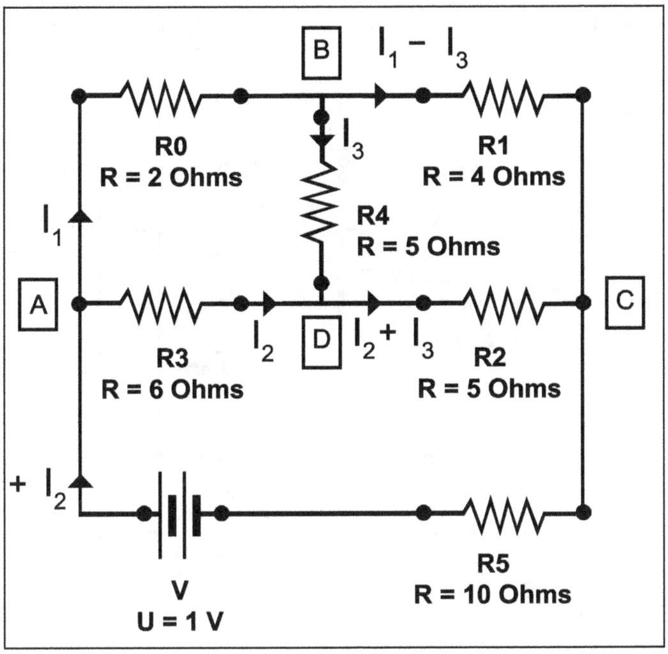

Loop ABDA:

$$-2I_1 - 8I_3 + 6I_2 = 0$$

$$\text{Or, } I_1 - 3I_2 + 4I_3 = 0$$

Loop BCDB:

$$-4(I_1 - I_3) + 5(I_2 + I_3) + 8I_3 = 0$$

$$\text{Or, } 4I_1 - 5I_2 - 17I_3 = 0$$

Loop ABCVA:

$$-2I_1 - 4(I_1 - I_3) - 10(I_1 + I_2) + 1 = 0$$

$$16I_1 + 10I_2 - 4I_3 = 1$$

Thus solving these three equations, we have:

$$I_1 = 0.0494\text{A}$$

$$I_2 = 0.0229\text{A}$$

$$I_3 = 0.0049\text{A}$$

Let us add an extra e.m.f of 1 volt on branch BD to oppose the flow of original current in that branch.

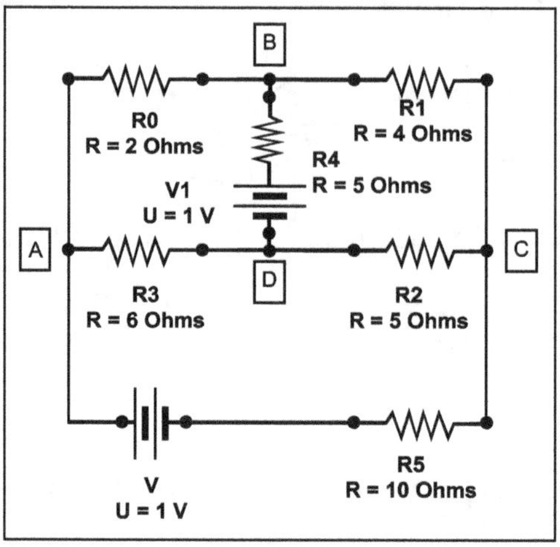

Now let us use the superposition theorem to find the current through the battery due to V and V1 separately.

Due to V

Current through battery due to $V = I_1 + I_3 = 0.0723\ A$

Due to V_1

Let us again use reciprocity theorem. The one voltage supply at V produces I_3 at branch BD. When -1 voltage acts on the branch BD, the current induced in the place of V is $-I_3$. So, current through V due to $V_1 = -I_3 = -0.0049\ A$

Current through V due to both voltages is given as:

$$0.0723\ A - 0.0049\ A = 0.0674\ A$$

4. Referring to the circuit shown in the below figure, let us determine the current through 10 Ω resistor using Millman's theorem.

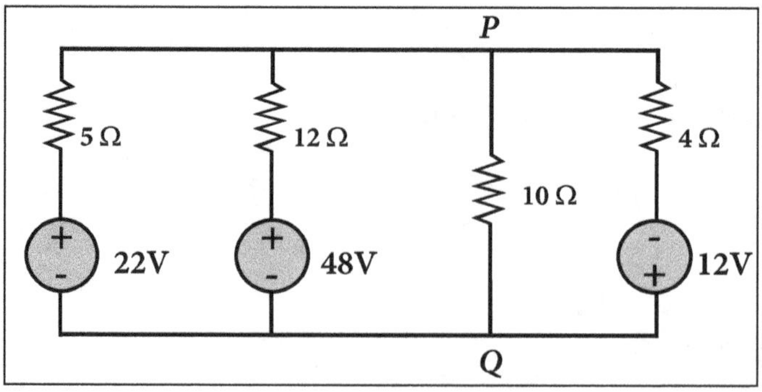

Solution:

Given:

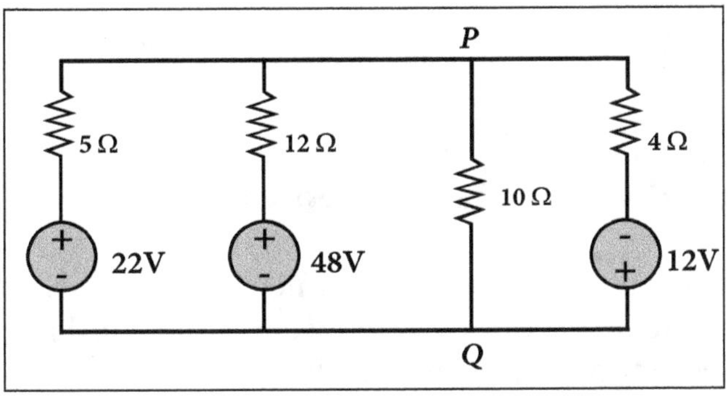

Using Millman's theorem, the circuit shown in the given figure is replaced by it's the venin equivalent circuit across the terminals PQ as shown in the below figure.

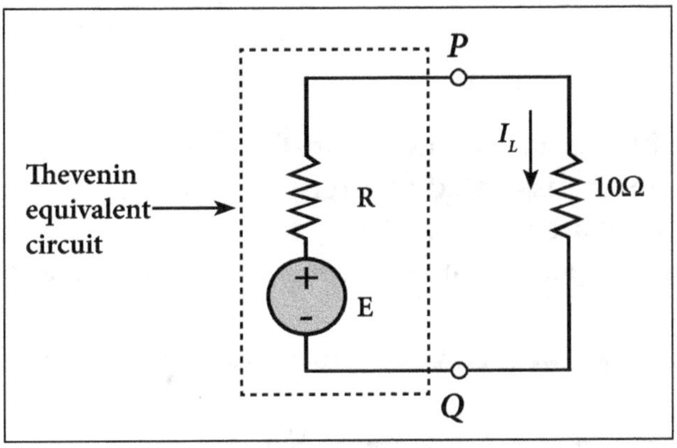

$$E = \frac{E_1 Y_1 + E_2 Y_2 - E_3 Y_3}{Y_1 + Y_2 + Y_3}$$

$$= \frac{22\left(\frac{1}{5}\right) + 48\left(\frac{1}{12}\right) - 12\left(\frac{1}{4}\right)}{\frac{1}{5} + \frac{1}{12} + \frac{1}{4}}$$

$$= 10.13 \text{ Volts}$$

$$R = \frac{1}{Y_1 + Y_2 + Y_3}$$

$$= \frac{1}{0.2 + 0.083 + 0.25}$$

$$=1.88\Omega$$

Hence,

$$I_L = \frac{E}{R+10} = 0.853 \text{ A}$$

2.1.6 Maximum Power Transfer Theorem

It is a theorem which is used in the electrical network analysis and electrical circuit designing. It was invented by a German engineer Moritz von Jacobi in the year 1840. He invented the theorem to maximize the output of the battery to a motored boat which he designed to travel in the river Neva. Hence, the theorem is also referred as Jacobi's law.

It deals with the power transferred to the load on a circuit with a network of various sources or components on it. It defines the condition under which the maximum power is transferred to the load in a circuit.

Under the condition of maximum power transfer, the power transferred to the load is considered and the power dissipated in internal circuits or resistance of the source is not considered. Thus, we are not considering the maximum efficiency of power transfer but instead maximum possible power transfer from a source to a load is taken into account.

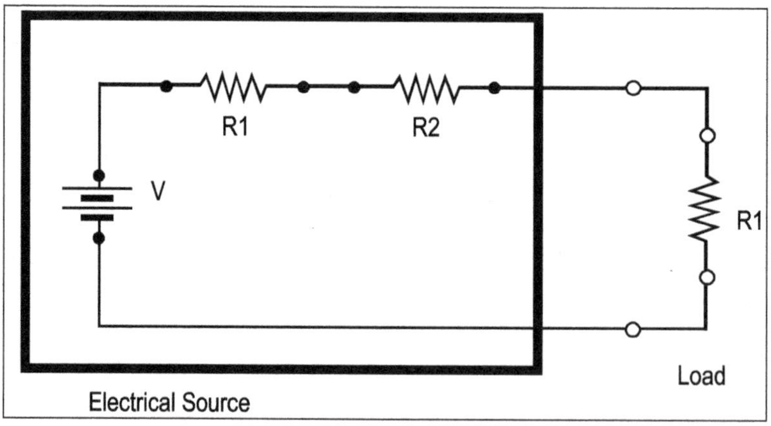

Example for maximum power transfer theorem.

According to maximum power transfer theorem, maximum power is yielded to the load Rl where RL is equal to the internal resistance of the circuit (R1 + R2). The maximum power transfer theorem is true for linear, non-linear, active, DC or AC circuits.

In DC circuits, load resistance is matched with the internal resistance of the source where both the resistances are equal and in AC; the load impedance is matched with the internal impedance of the circuit or source where the load impedance is the complex conjugate of the source impedance.

For example, if load impedance is R1 - jX, then the internal impedance of the source is R1+ jX.

Proof of Maximum Power Transfer Theorem

V = EMF supplied to the load

R_L = Load resistance

R_i = Internal resistance of the source

I = Current flowing through the load, internal resistance and the source of the circuit

P_L = Power transferred to the load

P_i = Power dissipated at the internal resistance

Power transferred to the load = $P_L = I^2R_L$

$$P_L = \left(\frac{V}{R_i + R_L}\right)^2 \times R_L = \frac{V^2}{\frac{R_i^2}{R_L} + 2R_i + R_L}$$

Using the theorems of differential calculus, if R_L is variable, then the maximum value of P_L is calculated. P_L is differentiated with respect to R_L and it is equated with zero. Thus, under maximum power transfer to load condition, we have:

$$\frac{d}{dR_L}P_L = \frac{d}{dR_l}\frac{V^2}{\frac{R_i^2}{R_L} + 2R_i + R_L} = 0$$

$$-\frac{R_i^2}{R_L^2} + 1 = 0$$

$$R_i = R_L$$

In AC networks, we have:

$Z_i = R_i + X_i$ = Internal impedance of reactive circuits

$Z_L = R_L + X_L$ = Load impedance

Under the condition of maximum power transfer to load, we have:

$$Z_i = Z_l$$

$$R_i = R_L$$

$$X_i = -X_L$$

Power Transfer Efficiency

It is the efficiency of any source or circuit in transferring its power to the load or it is the ratio of power transferred to the load over total power transferred by the source. It is denoted as η.

$$\eta = P_L / P_T$$

Where,

P_L = Power transferred to the load

P_T = Total power transferred by the source

$$P_L = I^2 R_L$$

$$P_T = P_L + P_I = I^2 R_L + I^2 R_I$$

Where,

I = Total current flowing through the circuit or Thevenin's equivalent of the circuit.

P_I = Power dissipated in internal circuits of the source.

Thus, $\eta = \frac{P_L}{P_T} = \frac{I^2 R_L}{I^2 R_L + I^2 R_I} = \frac{R_L}{R_L + R_I} = \frac{1}{1 + \frac{R_I}{R_L}}$

Under the condition of maximum power transfer:

$$R_L = R_I$$

Under the condition of maximum power transfer, efficiency is only 0.5.

The overall efficiency decreases if the value of R_L is very low and it increases up to one, when the value of R_L is increased to infinity as shown on the below graph and the power transferred to load is minimum when R_L is very low and very high.

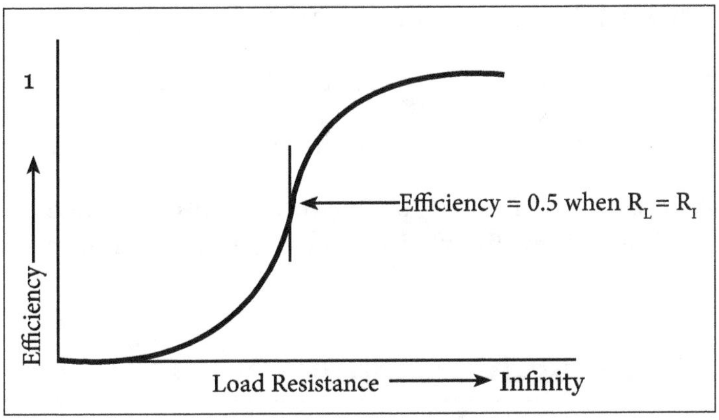

Power transfer efficiency vs load resistance.

Maximum power transfer to load is obtained when $R_L = R_I$ as shown on the below graph.

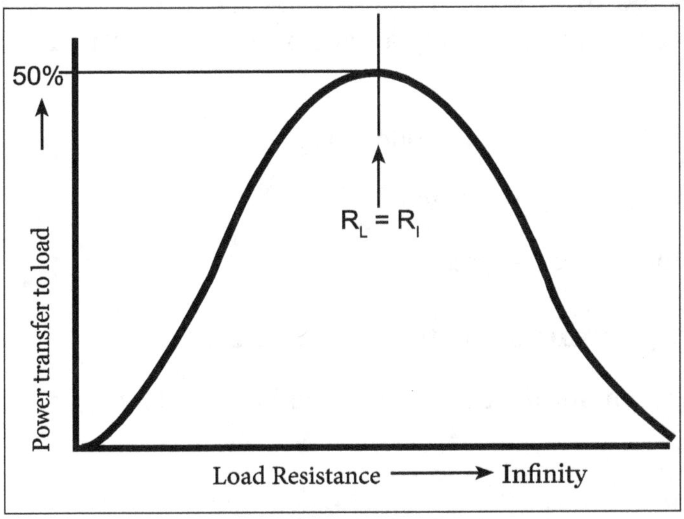

Power transfer to load vs load resistance.

Applications of Maximum Power Transfer Theorem

Maximum power transfer theorem has a wide range of usage on real life situation. This theorem is used to maximize the power output to a load from any circuit. So, it is used to design circuits where the maximum output performance is desired to match the amplifier with the loudspeaker to yield the maximum power to the speaker and it produces maximum sound.

Transformer coupling yields maximum power to the load when the matching of load and source impedance is not possible. The application of maximum power theorem is done only under the conditions where the maximum performance is desired over the overall efficiency of the circuit since the efficiency of the circuit under maximum power transfer condition is only 0.5.

2.1.7 Applications of Network Theorem

- It is used in the analysis of properties and behaviour of circuits.
- It is used to visualize the response of the complex network.
- Superposition theorem is used in the study of alternating current (AC) circuits and semiconductor (amplifier) circuits where AC is often mixed with DC.

2.2 Network Reduction: Voltage and Current Division

Network Reduced System

The nine-bus system is divided into an internal system and an external system. The internal system is the study area expected to be remained. In Version 13.2, the system is divided into two grids namely a distribution system and a transmission system.

New boundary is determined by the following steps.

- The network diagram is freezed.
- Desired cubicles are selected.

Technical Background of Network Reduction

The voltage distribution for the circuit is shown in the below figure.

$$V_1 = \left(\frac{R_1}{R_1 + R_2}\right) V$$

$$V_2 = \left(\frac{R_2}{R_1 + R_1}\right) V$$

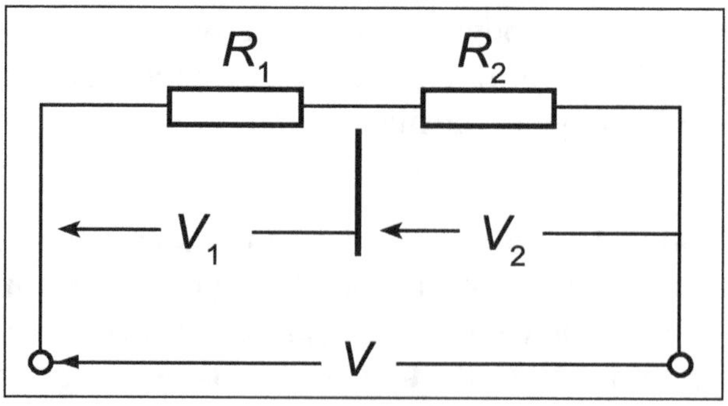

Voltage distribution.

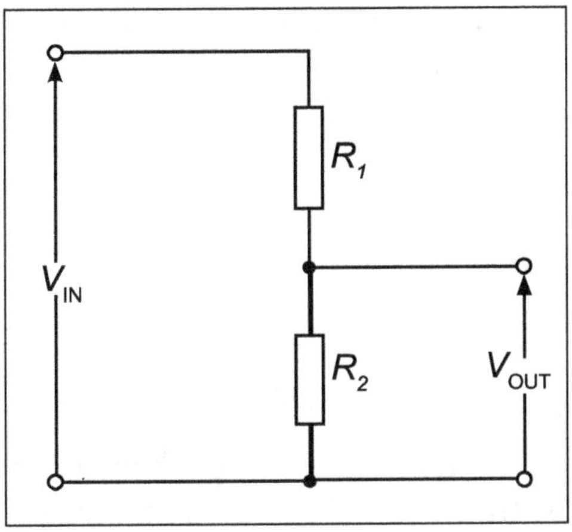

Potential divider circuit.

Potential divider circuit consists of a number of similar elements in series which are connected across a voltage source where voltages are taken from connections between the elements.

$$V_{OUT} = \left(\frac{R_2}{R_1 + R_2}\right) V_{IN}$$

A potential divider produces a source of lower e.m.f. from a source of higher e.m.f. and it is the basic operating mechanism of the potentiometer which is a measuring device for accurately measuring potential differences.

2.3 Source Transformation-Star Delta

Voltage and Current Source Transformation

The circuit of the below figure has 2 loops and 2 node pairs. The same can be transformed into a single node pair equivalent circuit so that the analysis can be done with ease without going through reduction of network and then establishing the division of voltages and currents.

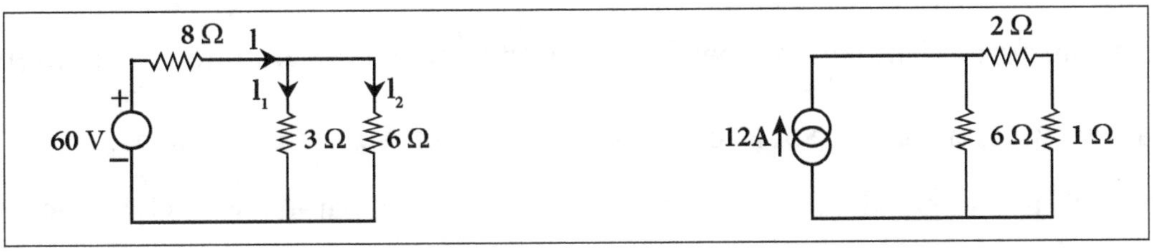

Similarly, the circuit of above figure having 2 node pairs and 2 loops can be transformed into a single loop circuit. The former is the transformation of an independent voltage source in series with the resistance into an independent current source in parallel with the conductance while the latter is the reverse.

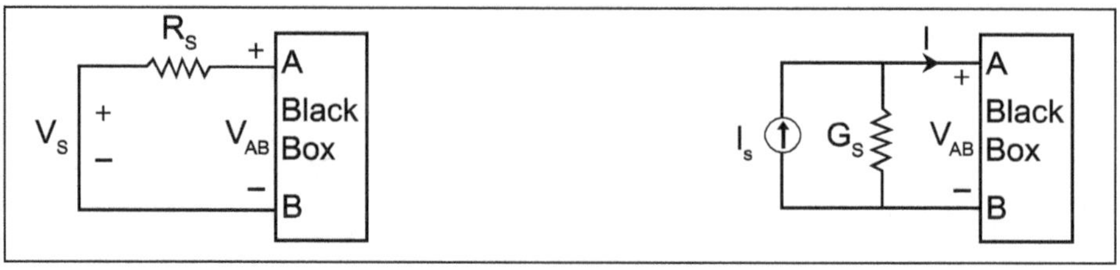

Voltage Source Transformation.

Consider the circuit shown in the above figure where in the black box represents a network. This circuit can be transformed into that (shown in the above figure), there is no change in terminal condition. That is, V_{AB}, the voltage across AB and the current I entering at A remains the same.

From the above figure:

$$I = \frac{V_S - V_{AB}}{R_S}$$

$$I = \frac{V_S}{R_S} - \frac{V_{AB}}{R_S} \qquad \text{...(1)}$$

From the above figure:

$$I = I_S - G_S V_{AB} \qquad \text{...(2)}$$

On comparison of (1) and (2)

$$I_S = \frac{V_S}{R_S}$$

and $$G_S = \frac{1}{R_S}$$

Therefore, an independent voltage source, V_S volts, in series with resistance RS ohms, can be transformed into an independent current source of magnitude $\left(\frac{V_S}{R_S}\right)$ amperes in parallel with a resistance of R_S ohms or a conductance of $(1/R_S)$ Siemens.

Conversely an independent current source of I amperes in parallel with G_S Siemens can be transformed into a voltage source of $\left(\frac{I_S}{G_S}\right)$ volts in series with $(1/G_S)$ Siemens or R_S ohms.

2.3.1 Star Delta

Transformation of a given set of resistance (or impedance) in star to delta or vice versa proves extremely useful in circuit analysis and the apparent complexity of a given network can sometimes be very much reduced.

Referring to the below figure suppose that resistances R_A, R_B and R_C are known and we require to find R_1, R_2 and R_3 such that the circuits are identical as for as terminals 1, 2, 3 are concerned. Consider the resistance between terminals 1 and 2 with terminals 3 floating:

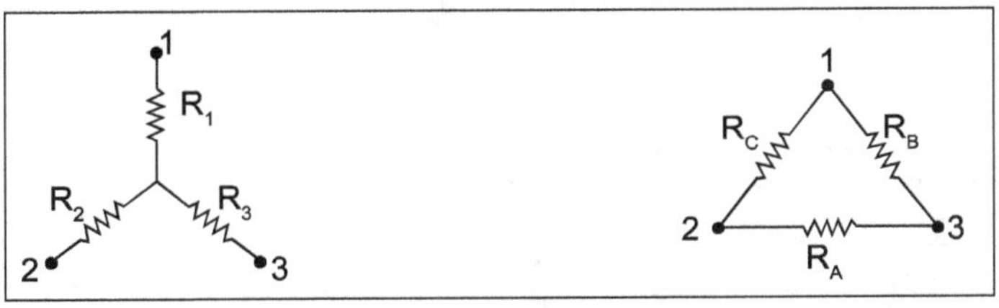

Illustrating Star delta Transformation.

$$R_{12} = R_1 + R_2 \quad ...(1)$$

$$R_{12} = \frac{R_C(R_A + R_B)}{R_A + R_B + R_C} \quad ...(2)$$

Similarly, resistance between terminals 2 and 3 with terminal 1 floating is:

$$R_{23} = R_2 + R_3 \quad ...(3)$$

$$R_{23} = \frac{R_A(R_B + R_C)}{R_A + R_B + R_C} \quad ...(4)$$

Resistance between terminals 3 and 1 with terminals 2 floating is:

$$R_{31} = R_3 + R_1$$

$$R_{31} = \frac{R_B(R_C + R_A)}{R_A + R_B + R_C} \quad ...(5)$$

Hence, solving the three equations:

$$\left.\begin{aligned} R_1 + R_2 &= \frac{R_C(R_A + R_B)}{\sum R} \rightarrow a \\ R_2 + R_3 &= \frac{R_A(R_B + R_C)}{\sum R} \rightarrow b \\ R_3 + R_1 &= \frac{R_B(R_C + R_A)}{\sum R} \rightarrow c \end{aligned}\right\} \quad ...(6)$$

Where, $\sum R = R_A + R_B + R_C$

Subtracting (c) and (b) we have:

$$R_1 - R_2 = \frac{R_C (R_B - R_A)}{\sum R} \quad \text{...(7)}$$

Adding the above equation (7) with (a) yields:

$$R_1 = \frac{R_B R_C}{\sum R}$$

Similarly, $R_2 = \dfrac{R_C R_A}{\sum R}$ and $R_3 = \dfrac{R_A R_B}{\sum R}$

To calculate the formulae for the reverse transformation i.e., the case where R_1, R_2, R_3 are known and we desire to find R_A, R_B, R_C we proceed as follows:

The conductance between terminals 1 and 2 with terminal 3 short-circuited to terminal 2 is:

$$G_{12} = \frac{G_1 (G_2 + G_3)}{G_1 + G_2 + G_3} = G_B + G_C \quad \text{...(8)}$$

Similarly between 2 and 3 with 1 shorted to 3 is:

$$G_{23} = \frac{G_2 (G_3 + G_1)}{G_1 + G_2 + G_3} = G_C + G_A \quad \text{...(9)}$$

And the conductance between 3 and 1 with 2 shorted to 1 is:

$$G_{31} = \frac{G_3 (G_1 + G_2)}{G_1 + G_2 + G_3} = G_A + G_B \quad \text{...(10)}$$

Solving the above equations (8), (9) and (10):

$$G_A = \frac{G_2 G_3}{\Sigma G}$$

$$G_B = \frac{G_3 G_1}{\Sigma G}$$

$$G_C = \frac{G_1 G_2}{\Sigma G}$$

Where, $G=G_1+G_2+G_3$

Problems

1. Let us consider the following network in the below figure and let us also determine the power absorbed by 6 ohms resistor.

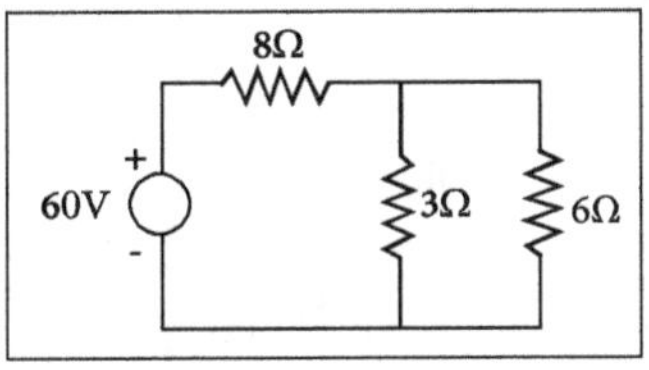

Solution:

$$\frac{1}{R_{eq}}=\frac{1}{8}+\frac{1}{3}$$

$$\frac{1}{R_{eq}}=\frac{3+8}{24}=\frac{11}{24}$$

$$R_{eq}=\frac{24}{11}$$

$$I_{6\Omega}=\frac{\frac{60}{8}\times\frac{24}{11}}{\frac{24}{11}+6}=2\,A$$

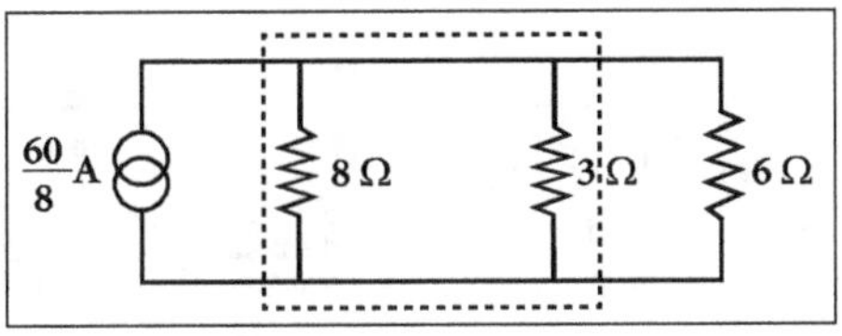

Power absorbed in 6 Ω resistor = (2)2 x 6 = 24 Watts.

2. Let us find the value of a single resistance to replace the network between terminals A and B of the below given figure.

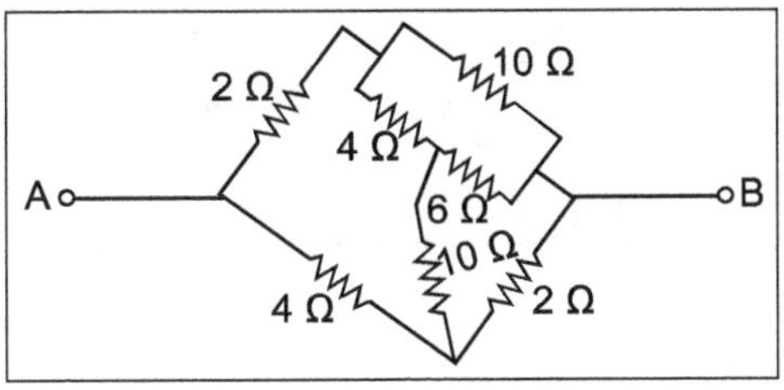

Solution:

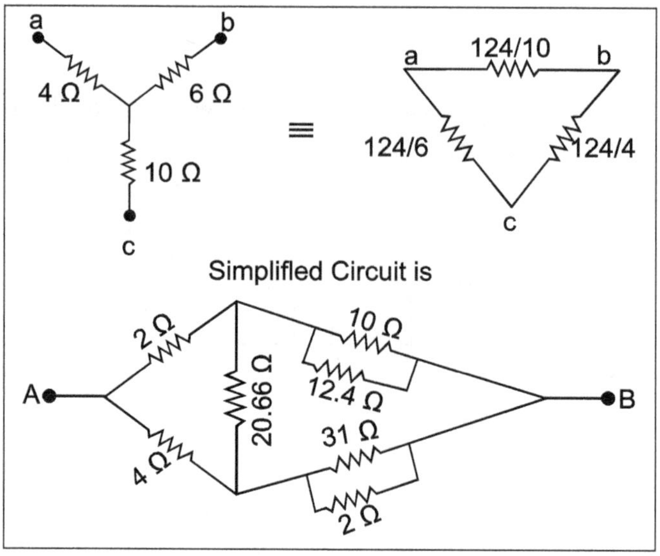

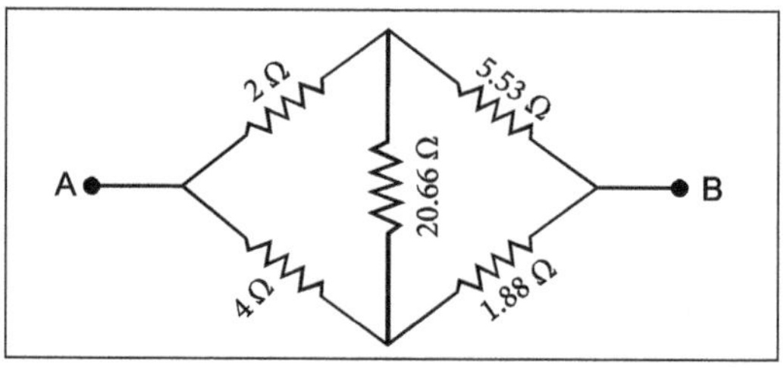

$2\Omega \| 31\Omega = 1.88\ \Omega$

$10\Omega \| 12.4\Omega = 5.53\ \Omega$

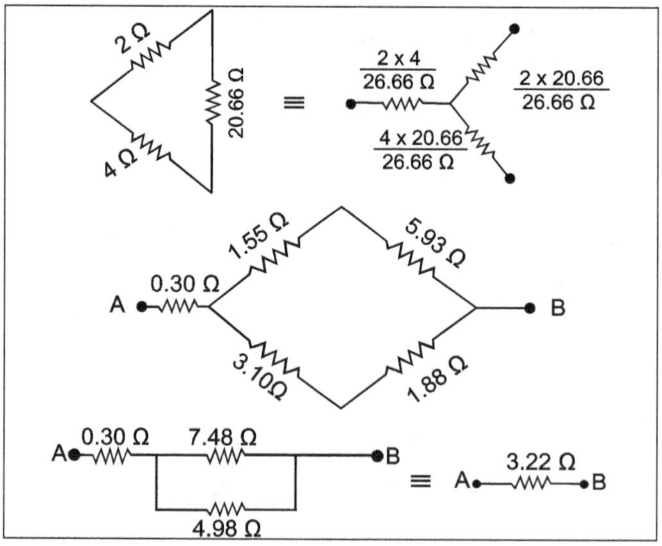

3. Let us obtain the equivalent resistance RAB in the circuit of below figure.

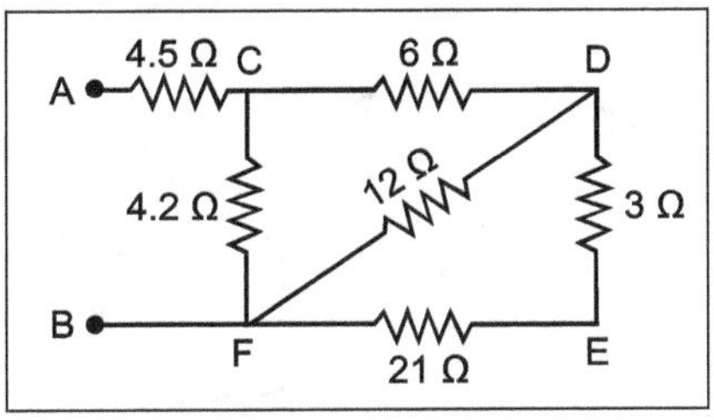

Solution:

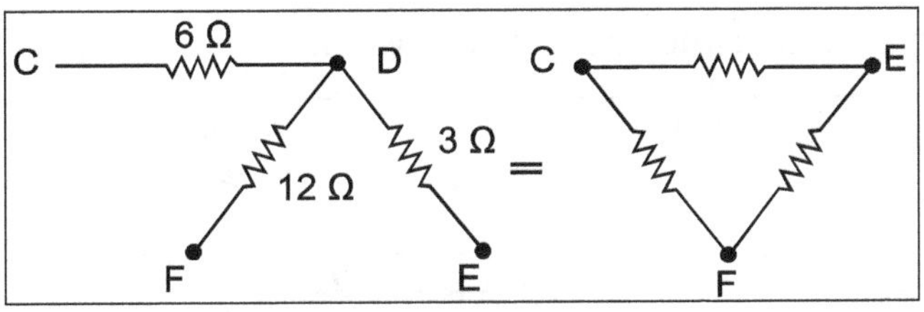

$$R_{CE} = \frac{(6\times3)+(3\times12)+(12\times6)}{12} = 10.5\ \Omega$$

$$R_{EF} = \frac{(6\times3)+(3\times12)+(12\times6)}{6} = 21\ \Omega$$

$$R_{FC} = \frac{(6\times3)+(3\times12)+(12\times6)}{3} = 42\ \Omega$$

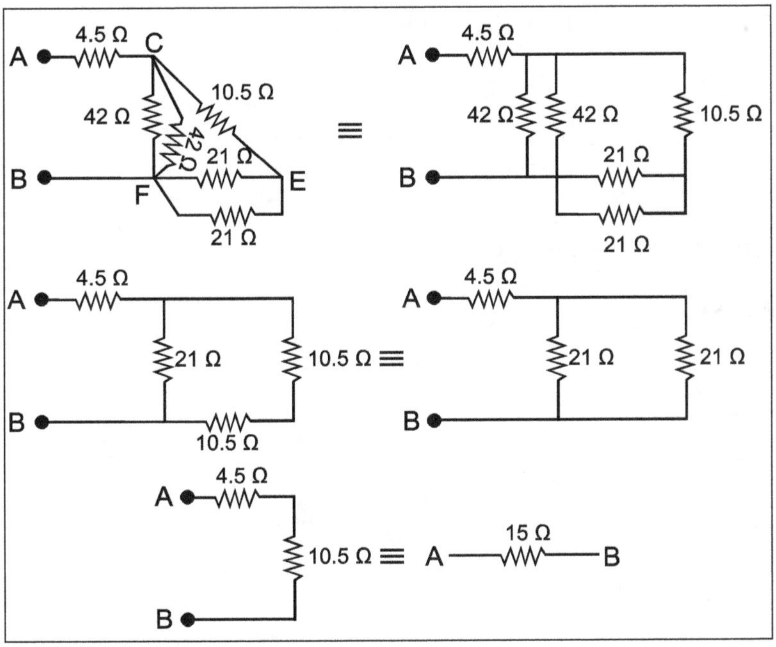

4. Let us determine the current through 16Ω resistor for the below figure.

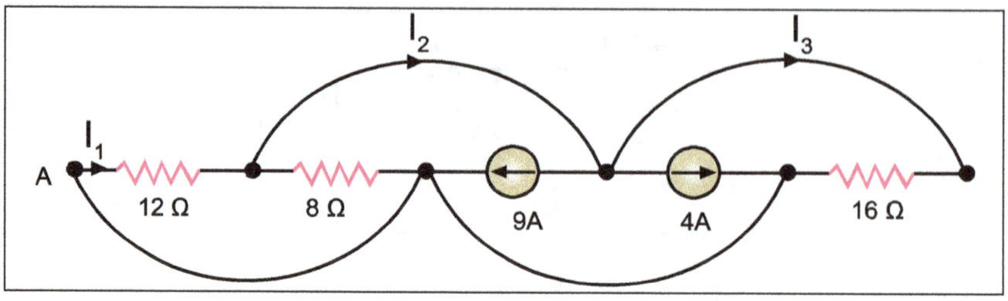

Solution:

Assigning currents to each branch using KCL, we have:

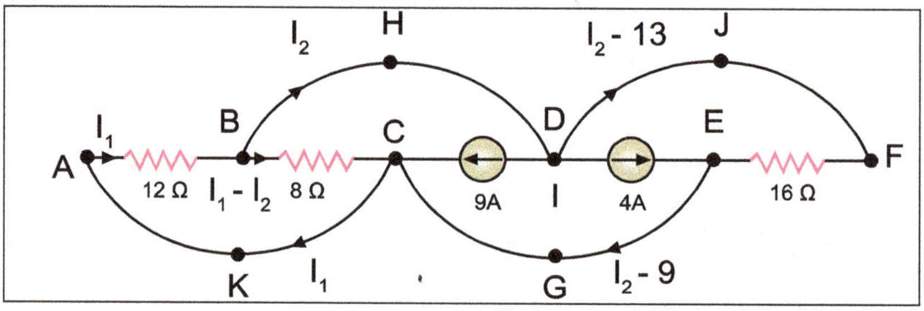

By applying KVL to ABCKA, we have:

$$-12\,I_1 - 8(I_1 - I_2) = 0$$

$$-20\,I_1 + 8\,I_2 = 0 \qquad ...(1)$$

By applying KVL to ABHDJFEGCKA, we have:

$$-12\,I_1 - 16(I_2 - 1_3) = 0$$

$$-12\,I_1 - 16\,I_2 = -208 \qquad ...(2)$$

Solving equations (1) and (2), we get:

$$I_1 = 4\,A(\rightarrow)$$

$$I_2 = 10\,A\,(\curvearrowright)$$

$$I_{16} \;= I_2 - I_3$$

$$= 10 - 13 = -3A$$

$$I_{16\,\Omega} = 3A(\rightarrow)$$

5. Let us determine the value of voltage V shown in the below figure.

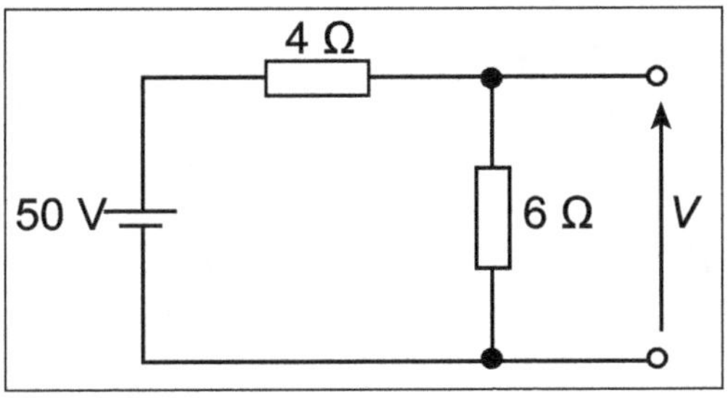

Solution:

$$V = \left(\frac{6}{6+4}\right)(50) = 30\,V$$

The resultant figure is given as:

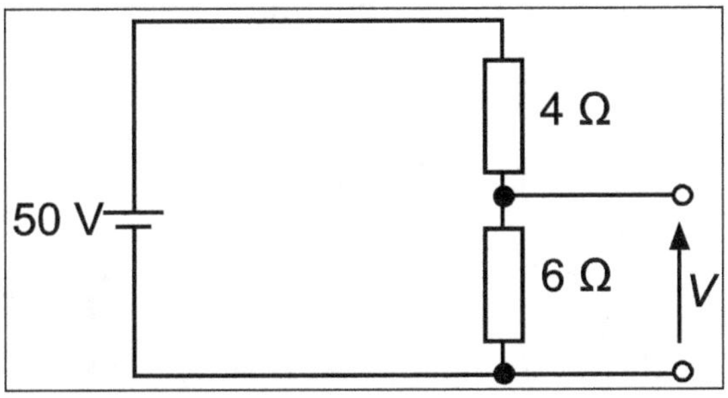

6. Let us consider two resistors which are connected in series across a 24V supply and a current of 3A flows in the circuit. If one of the resistors has resistances of 2 Ω, let us determine the following,

(a) Value of the other resistor.

(b) Power dissipation across the 2 Ω resistors.

If the circuit is connected for 50 hours, let us calculate the amount of energy.

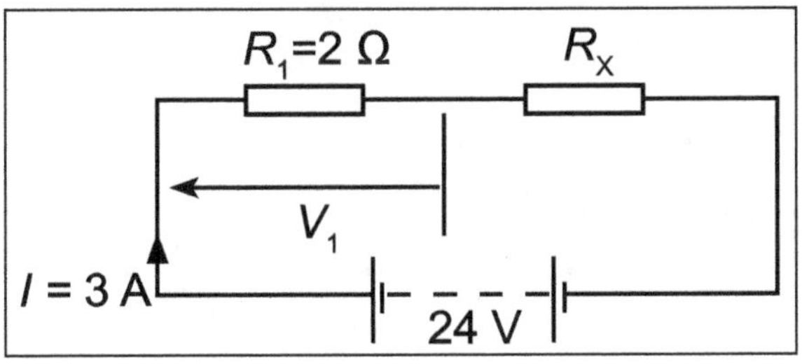

Solution:

Total circuit resistance, $R = V/I = 24/3 = 8\,\Omega$

Value of unknown resistance, $R_x = 8 - 2 = 6\ \Omega$

(b) Power dissipation across 2 Ω resistors:

$$V_1 = IR_1 = 3 \times 2 = 6V$$

$$V_1 = (R_1 / R_1 + R_x))\ V = (2/2+6)(24) = 6V$$

$$= V \times I \times t$$

$$= (24 \times 3W)(50h)$$

$$= 3600\,Wh = 3.6kWh$$

Current Division Rule

For the circuit shown in the below figure, the total circuit resistance (R_T) is given as:

$$R_T = R_1R_2 / R_1 + R_2$$

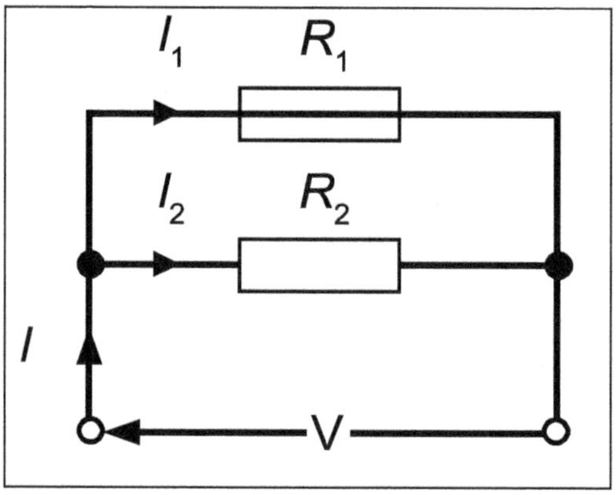

$$V = IR_T = I\left(\frac{R_1R_2}{R_1+R_2}\right)$$

I_1 is given as:

$$I_1 = \frac{V}{R_1} = \frac{I}{R_1}\left(\frac{R_1R_2}{R_1+R_2}\right) = \left(\frac{R_2}{R_1+R_2}\right)(I)$$

I_2 is given as:

$$I_2 = \frac{V}{R_2} = \frac{I}{R_2}\left(\frac{R_1R_2}{R_1+R_2}\right) = \left(\frac{R_1}{R_1+R_2}\right)(I)$$

The total current of I_1 and I_2 is given as:

$$I_1 = \left(\frac{R_2}{R_1 + R_2}\right)(I) \quad \text{and} \quad I_2 = \left(\frac{R_1}{R_1 + R_2}\right)(I)$$

1. From the series-parallel arrangement shown in the below figure, let us find:

(a) Supply current.

(b) Current flowing through each resistor.

(c) Power dissipation across each resistor.

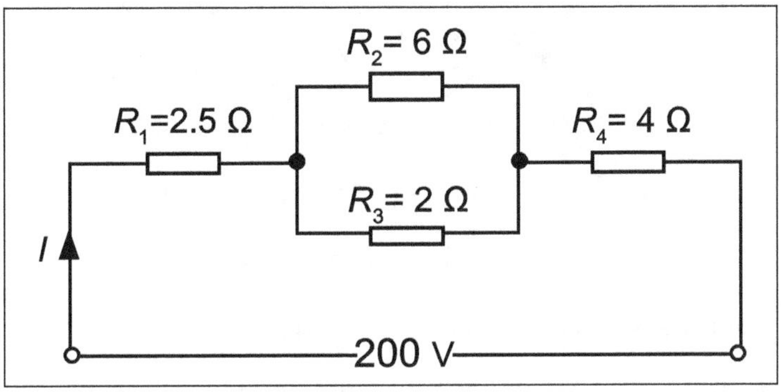

Solution:

(a) The equivalent resistance (R_x) of R_2 and R_3 in parallel is given as:

$$R_x = 6 \times 2 / 6 + 2$$

$$= 12 / 8$$

$$= 1.5\ \Omega$$

The equivalent resistance (R_T) of R_1, R_x and R_4 in series is given as:

$$R_T = 2.5 + 1.5 + 4 = 8\ \Omega$$

Supply current is given as:

$$I = V / R_T$$

$$= 200 / 8$$

$$= 25A$$

(b) Current flowing through R_1 and R_4 is 25A

Current flowing through R_2 is given as:

$$=\left(\frac{R_3}{R_2+R_3}\right)I=\left(\frac{2}{6+2}\right)25$$

$$=6.25A$$

Current flowing through R_3 is given as:

$$=\left(\frac{R_2}{R_2+R_3}\right)I=\left(\frac{6}{6+2}\right)25$$

$$=18.75A$$

(c) Equivalent circuit of the below figure is given as:

Power dissipation across R_1, i.e. $V_1=IR_1=(25)(2.5)=62.5V$

Power dissipation across R_x, i.e. $V_x=IR_x=(25)(1.5)=37.5V$

Power dissipation across R_4, i.e. $V_4=IR_4=(25)(4)=100V$

Power dissipation across R_2 = Power dissipation across R_3 = 37.5V

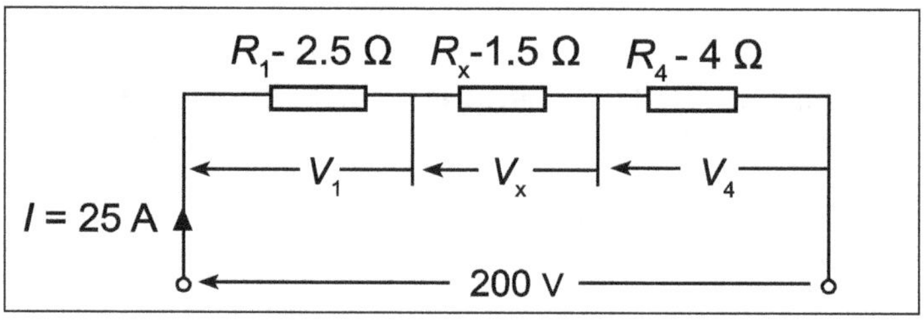

2. If an 8Ω loudspeaker is connected to an amplifier with an output impedance of 1000Ω, let us calculate the turns ratio of the matching transformer which will provide maximum power transfer of an audio signal. Let us assume that the amplifier source impedance is Z_1, the load impedance is Z_2 where the turns ratio is N.

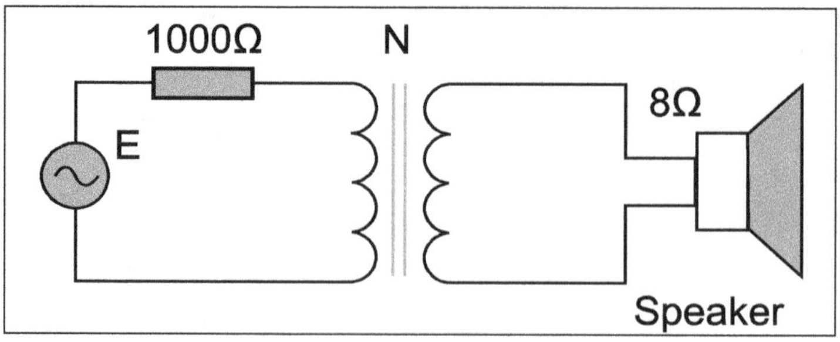

Transformer impedance matching circuit.

Solution:

Given:

8Ω loudspeaker:

Output impedance = 1000Ω

Source impedance of an amplifier = Z_1

Load impedance = Z_2

Transformer turns ratio is given as,

$$Z_1 = N\ Z_2$$

$$\therefore N = \sqrt{\frac{Z_1}{Z_2}}$$

$$N = \sqrt{\frac{Z_1}{Z_2}} = \sqrt{\frac{1000}{8}} = 11.2:1$$

Small transformers which are used in low power audio amplifiers are usually regarded as ideal. So, any losses can be ignored.

3. Let us determine the value of R_L for the given network where power is maximum and let us also determine the maximum power through load-resistance (R_L) by using maximum power transfer theorem.

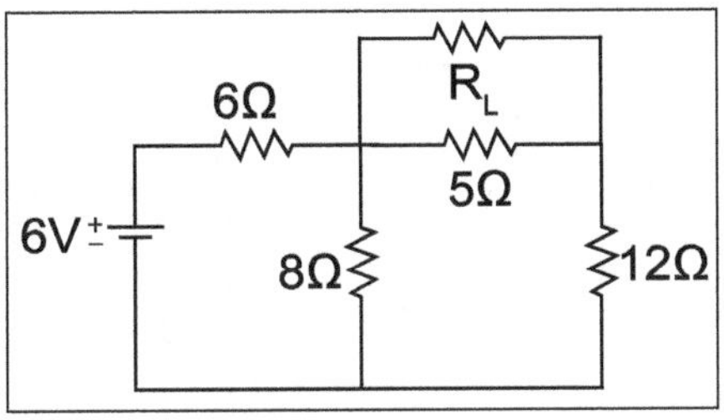

Solution:

Given:

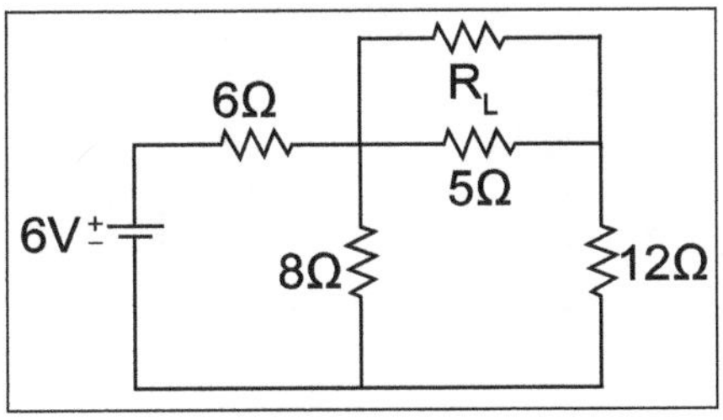

For the above network, let us determine the value of unknown resistance (R_L). When power is maximum through the load, resistance is equal to the equivalent resistance between the two ends of load-resistance. So, for finding load resistance (R_L), let us determine the equivalent resistance for the above circuit.

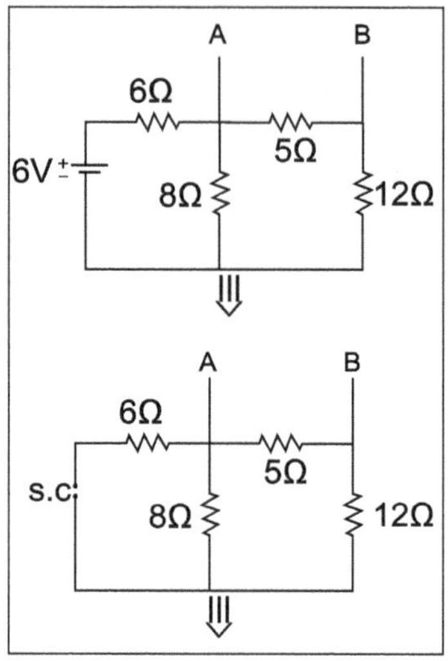

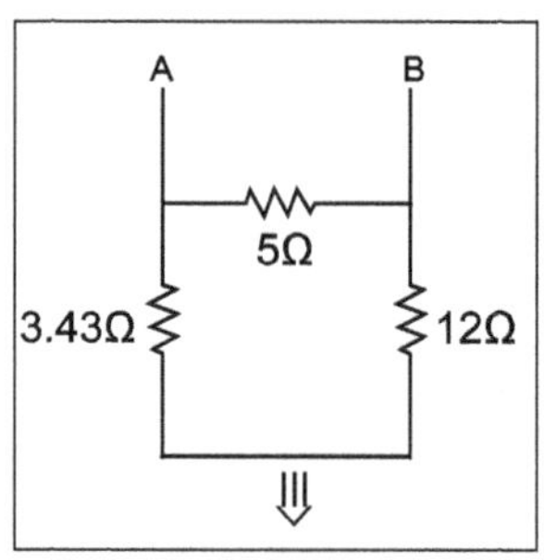

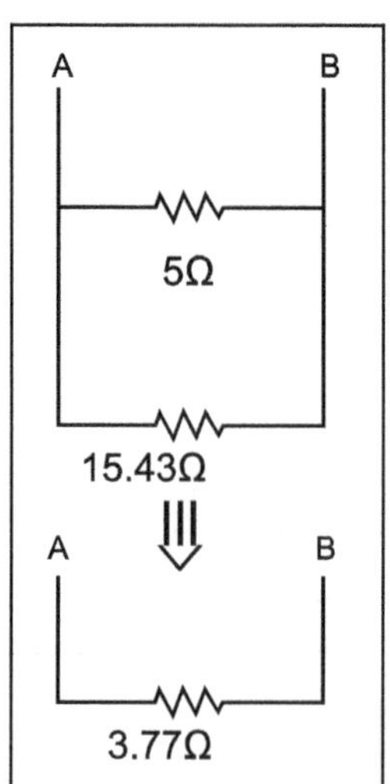

$$R_{AB} / R_L = 3.77\Omega$$

To finding maximum power through load-resistance, let us find-out the value of $V_{o.c}$.

Here, $V_{o.c}$ is known as the voltage between open circuits.

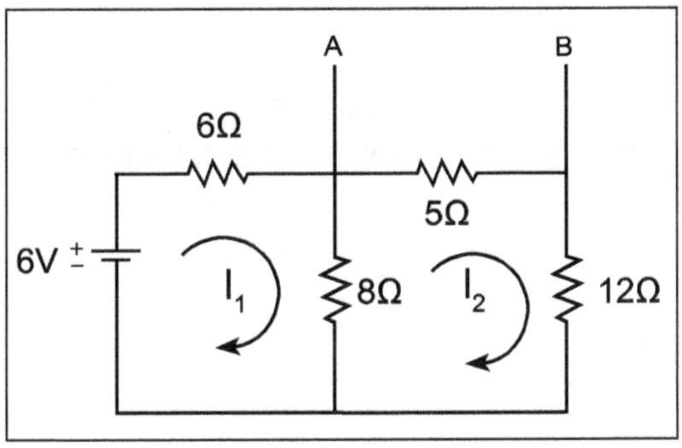

Applying KVL in 1st loop, we get:

$$6-6I_1-8I_1+8I_2=0$$

$$-14I_1+8I_2=-6 \quad ...(1)$$

Applying KVL in 2nd loop, we get:

$$-8I_2-5I_2-12I_2+8I_1=0$$

$$8I_1-25I_2=0 \quad ...(2)$$

On solving equations (1) and (2), we get:

$$I_1=0.524\text{ A}$$

$$I_2=0.167\text{ A}$$

From the circuit $V_{o.c}$, we have:

$$V_A-5I_2-V_B=0$$

$$V_{o.c}/V_{AB}=5I_2$$

$$=5\times0.167$$

$$=0.835\text{V}$$

Maximum power through R_L is given as:

$$P_{max}=\frac{V_{o.c}^2}{4R_L}$$

$$P_{max} = \frac{0.835^2}{4 \times 3.77}$$

$$P_{max} = 0.046 \text{Watt}$$

4. The effective resistance of two resistors connected in series is 100 Ω. When connected in parallel, then effective value in 24 ohms. Let us determine the value of two resistors.

Solution:

Given:

$$R_1 + R_2 = 100\ \Omega$$

$$R_1R_2 / R_1 + R_2 = 24$$

Series resistance is given as:

$$R_1 + R_2 = 100\ \Omega$$

$$=> R_2 = 100 - R_1$$

$$R_1R_2 / R_1 + R_2 = 24$$

$$R_1R_2 / 100 = 24$$

$$R_1R_2 = 2400$$

$$R_1(100 - R_1) = 2400$$

$$100\ R_1 - R_1^{\ 2} - 2400 = 0$$

$$R_1^{\ 2} - 100R_1 + 2400 = 0$$

$$(R_1 - 60)(R_1 - 40) = 0$$

$$R_1 = 60,\ R_1 = 40$$

When $R_1 = 60\ \Omega$,

$$R_2 = 100 - 60 = 40\ \Omega$$

When $R_1 = 40\ \Omega$,

$$R_2 = 100 - 40 = 60\ \Omega$$

5. Let us determine the R_{eq} between two points A & B.

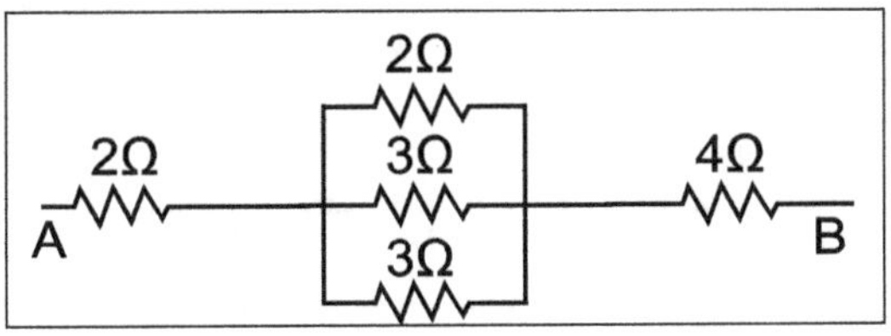

Solution:

Given:

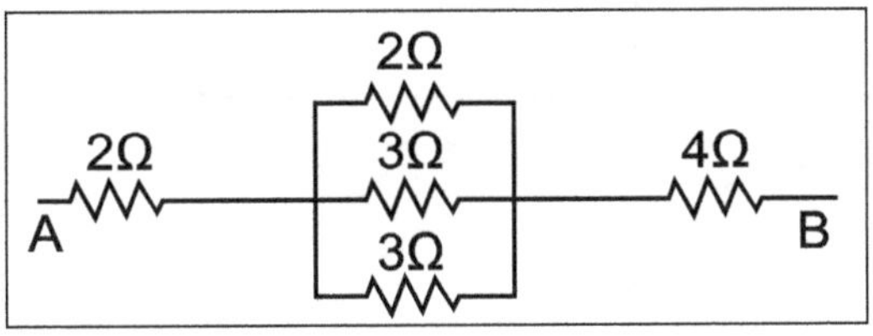

$R_1 = 2\Omega$

$R_2 = 3\Omega$

$R_3 = 3\Omega$

Formula to be used:

$$1/R_{eq} = 1/R_1 + 1/R_2 + 1/R_3$$

$$1/R_{eq} = 1/2 + 1/3 + 1/3 = 1.17$$

$$R_{eq} = 1/1.17 = 0.8547$$

$$R_{eq} = 0.854\Omega$$

6. A 47μF capacitor is connected to a voltage which varies in time as v(t) = 20sin(200πt) V. Let us calculate the current i(t) through the capacitor.

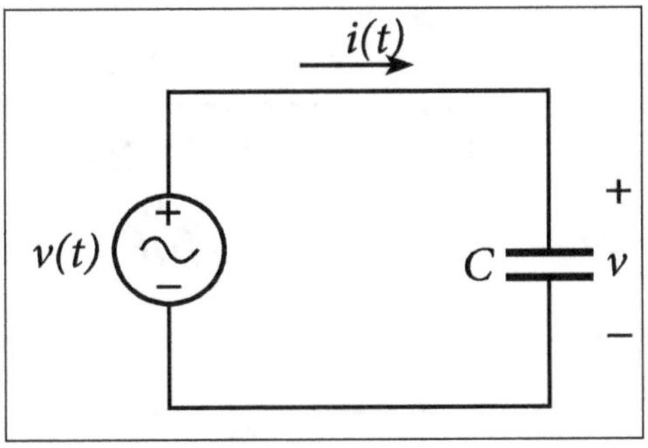

Solution:

Given:

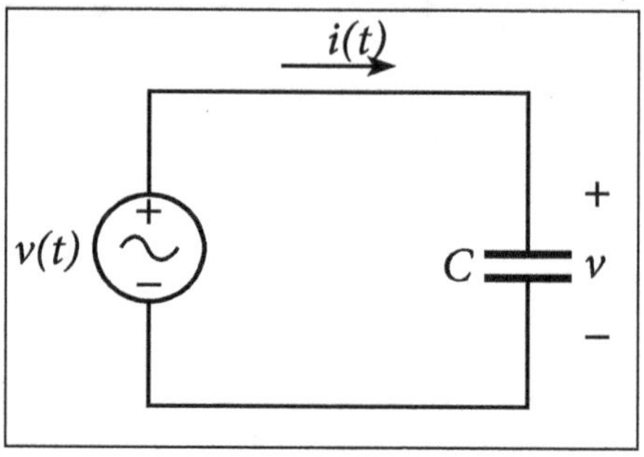

$$C = 47\mu F$$

$$V(t) = 20 \sin(200\pi t) \text{ volts}$$

Formula

$$i = C\frac{dv}{dt}$$

Current is given as:

$$i = C\frac{dv}{dt}$$

$$= 47\times10^{-6}\frac{d}{dt}20\sin(200\pi t)$$

$$= 47\times10^{-6}\times 20\times200\ \pi\cos(200\pi t)$$

$$= 0.59\cos(200\pi t) \text{ Amperes}$$

7. Let us calculate the energy stored in the capacitor of the circuit to the right under DC conditions.

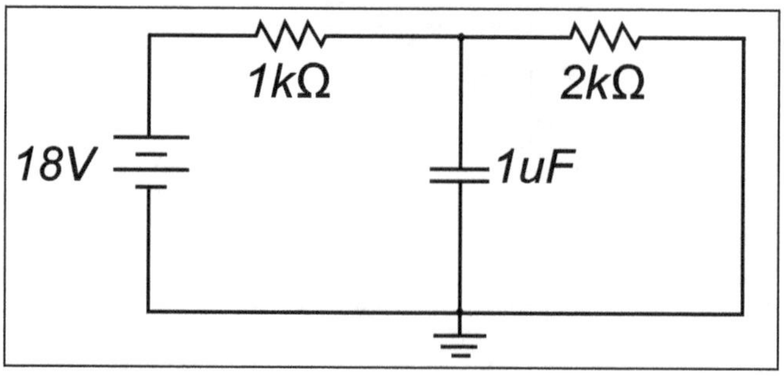

Solution:

Given:

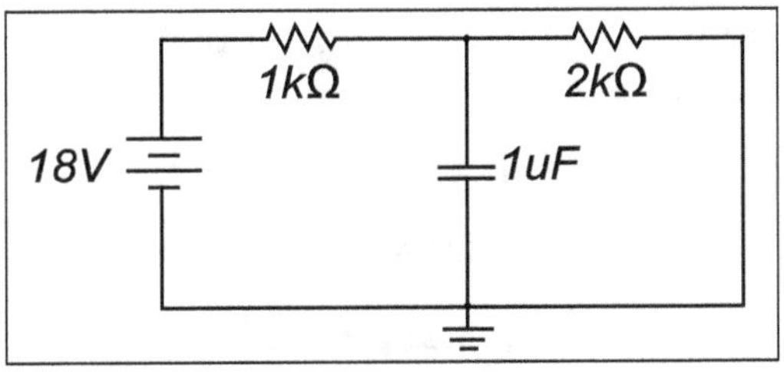

In order to calculate the energy stored in the capacitor, let us determine the voltage across it. Under DC conditions, the capacitor appears as an open circuit (no current flows through it). Thus, the corresponding circuit is shown in the below figure.

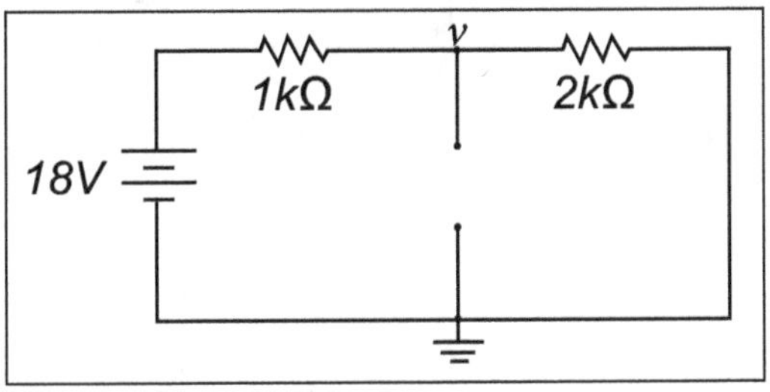

From the voltage divider formed by the 1kΩ and the 2kΩ resistors, the voltage (V) is 12Volts. Thus, the energy stored in the capacitor is given as:

$$Ec = \frac{1}{2}Cv^2 = \frac{1}{2}1\times10^{-6} \times 12^2 = 72\mu \text{Joules}$$

8. A 415 V, 3-phase, 4 wire, star-connected system supplies three resistive loads as shown in the below figure. Let us determine the following,

(a) Current in each line.

(b) Current in the neutral conductor.

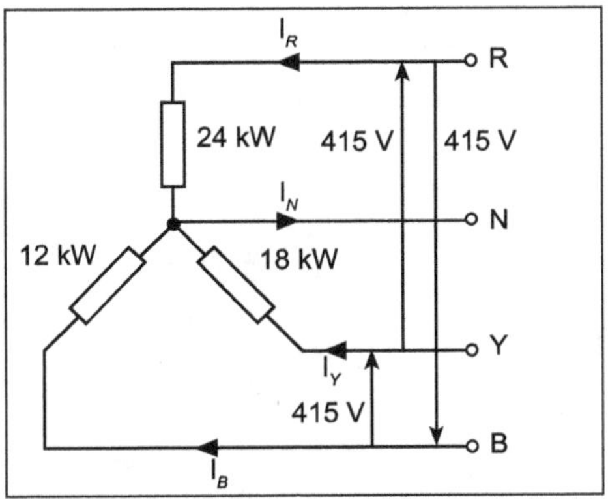

Solution:

Given:

A 415V, 3-phase, 4 wire, star-connected system supplies three resistive loads.

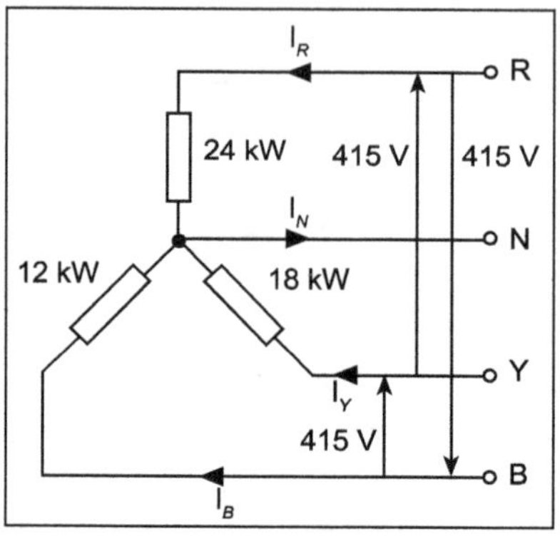

(a) For a star-connected system.

$$V_L = \sqrt{3}V_p$$

$$V_P = \frac{V_L}{\sqrt{3}} = \frac{415}{\sqrt{3}} = 240\,V$$

Current is given as:

$$I = \text{Power }(P) / \text{Voltage}(V)$$

Then:

$$I_R = P_R / V_R = 24\,000/240 = 100A$$

$$I_Y = P_Y / V_Y = 18\,000/240 = 75A$$

$$I_B = P_B / V_B = 12\,000/240 = 50A$$

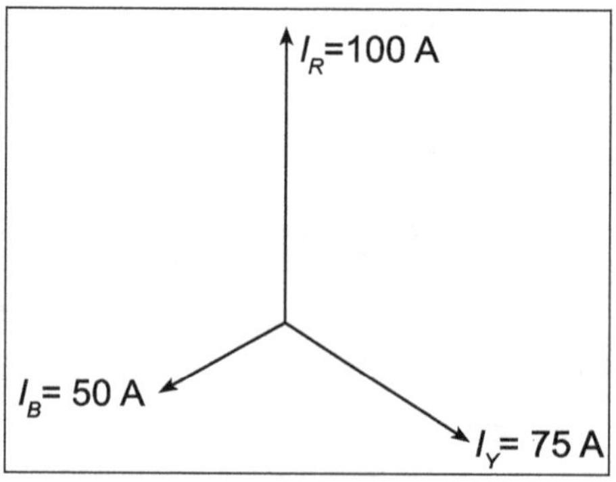

(b) The three line currents are shown in the below phasor diagram. Since each load is resistive, the currents are in phase with the phase voltages and they are mutually displaced by 120°.

The current in the neutral conductor is given as:

$$I_N = I_R + I_Y + I_B \text{ (Phasorially)}$$

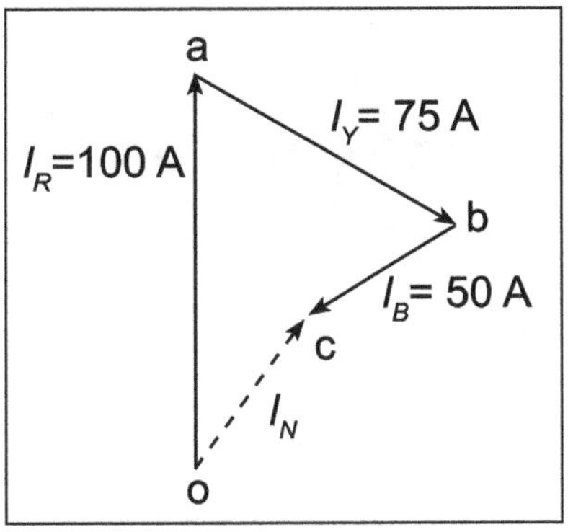

The above figure shows the three line currents which are added phasorially. oa represents IR in magnitude and direction. From the nose of oa, ab is drawn which represents IY in both magnitude and direction. From the nose of ab, bc is drawn which represents IB in both magnitude and direction and oc represents the resultant current, I_N.

By measurement:

I_N = 43A

Since I_R is at 90°, I_B is at 210° and I_Y is at 330°, we have,

Total horizontal component = 100 cos 90° + 75 cos 330°+50 cos 210° = 21.65

Total vertical component = 100 sin 90°+ 75 sin 330°+50 sin 210° = 37.50

Magnitude of I_N = $\sqrt{}$ (21.65² + 37.50²) = 43.3A

9. Let us convert the given delta figure into equivalent star.

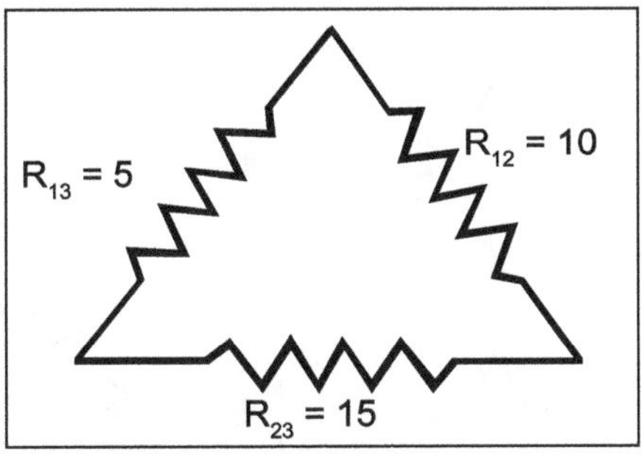

Solution:

Given:

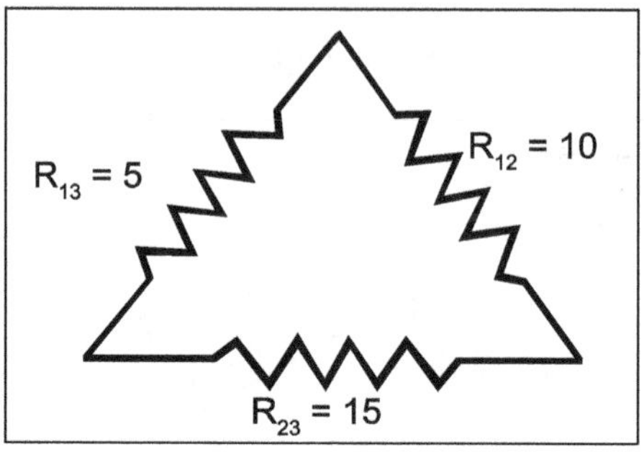

$$R_1 = R_{12} \times R_{13} / R_{12} + R_{13} + R_{23}$$

$$R_1 = 10 \times 5 / 30 = 1.67\Omega$$

$$R_2 = 10 \times 15 / 30 = 5\Omega$$

$$R_3 = 5 \times 15 / 30 = 2.5\Omega$$

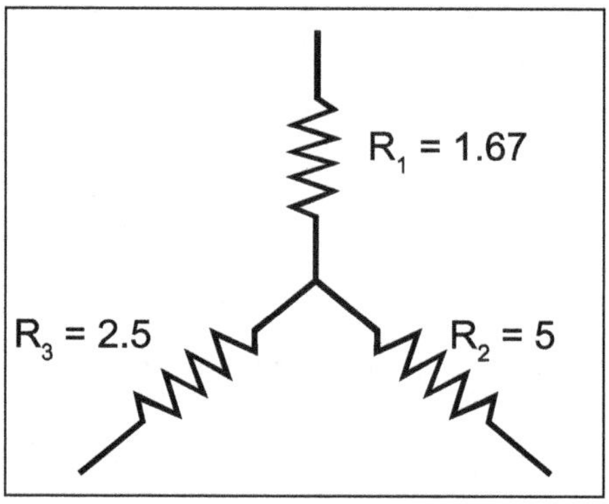

10. Let us consider that the below figure transform delta to star or wye circuit where the R_1 = 20 ohms, R_2 = 30 ohms and R_3 = 50 ohms. Let us determine the value of R_a, R_b and R_c.

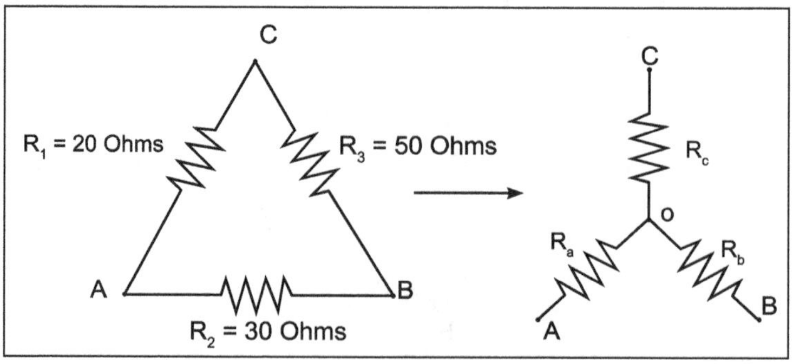

Solution:

Given:

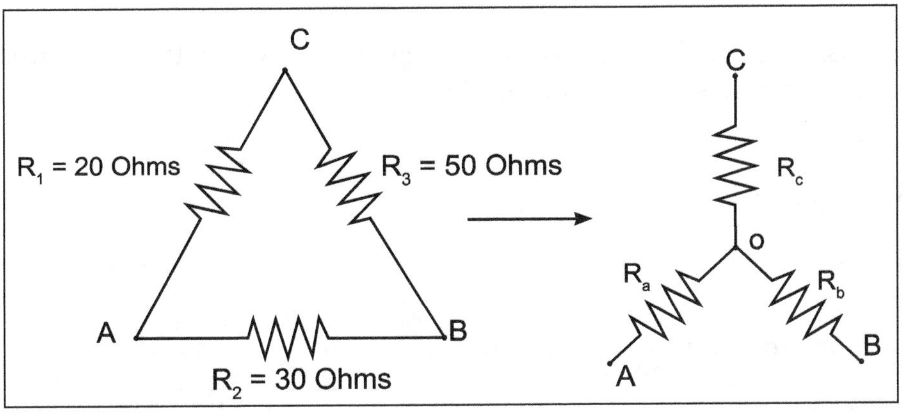

For delta to star conversions, equivalent resistance equations are given as:

$$R_a = (R_1R_2)/(R_1 + R_2 + R_3)$$

$$R_b = (R_2R_3)/(R_1 + R_2 + R_3)$$

$$R_c = (R_1R_3)/(R_1 + R_2 + R_3)$$

Total resistance is given as:

$$R_t = (R_1 + R_2 + R_3)$$

$$= 20 + 30 + 50$$

$$= 100\Omega$$

$$R_a = (R_1R_2)/(R_1 + R_2 + R_3)$$

$$= (20 \times 30)/100$$

$$= 6\Omega$$

$$R_b = (R_2R_3)/(R_1 + R_2 + R_3)$$

$$= (30 \times 50)/100$$

$$= 15\Omega$$

$$R_c = (R_1\ R_3)/(R_1 + R_2 + R_3)$$

$=(50\times 20)/100$

$=10\Omega$

11. Let us consider the below figure to transform star or wye to the delta circuit where the resistance values in star network are given as:

$R_1 = 10\Omega$

$R_2 = 5\Omega$

$R_3 = 20\Omega$

Let us determine the value of R_{12}, R_{23} and R_{31}.

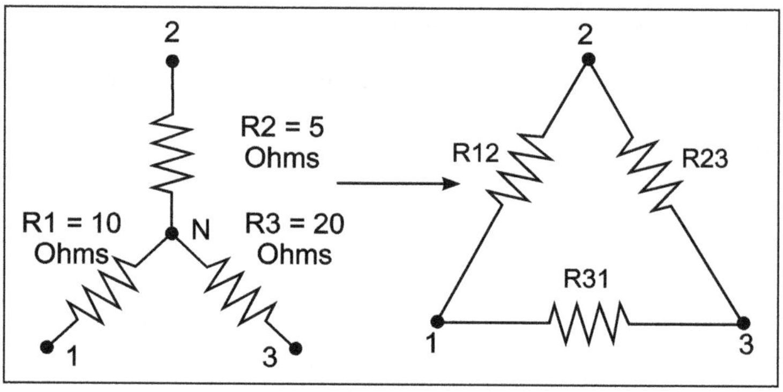

Solution:

Given:

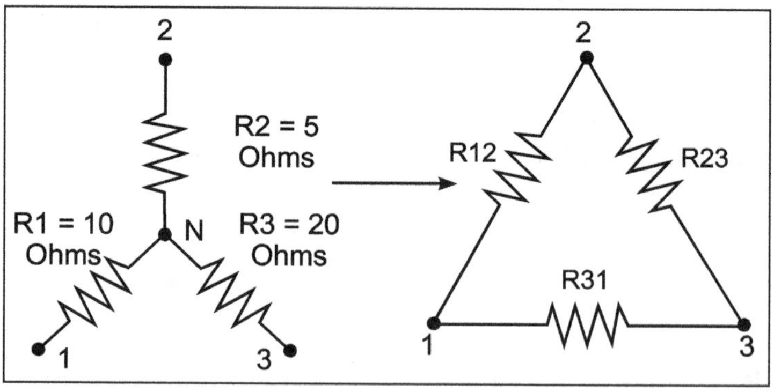

For star to delta conversion, the equivalent resistance equations are as follows:

$R12 = R1 + R2 + ((R1R2)/R3)$

$R23 = R2 + R3 + ((R2R3)/R1)$

$$R31 = R1 + R3 + ((R1R3)/R2)$$

By simplifying the above equations, we get the common numerator term as follows:

$$R1R2 + R2R3 + R1R3 = (10 \times 5) + (10 \times 20) + (20 \times 5)$$

$$= 350\Omega$$

$$R12 = 350/R3$$

$$= 350/20$$

$$= 17.5\Omega$$

$$R23 = 350/R1$$

$$= 350/10$$

$$= 35\Omega$$

$$R31 = 350/R2$$

$$= 350/5$$

$$= 70\Omega$$

3

Resonance and Coupled Circuits

3.1 Series and Parallel Resonance

Resonance Condition

When $X_L = X_C$ or the current in the circuit is in-phase with the applied voltage, then it is called as resonance condition. Imaginary part of the impedance (Z) or admittance (Y) is equated to zero to get resonance condition [i.e., if $Z = Re + jIm$, then $Im = 0$].

Types of resonance circuits:

- Series resonance.
- Parallel resonance.

Series Resonance

The basic series-resonant circuit is shown in the below figure. Since the frequency of the source changes, the maximum amplitude of the source voltage (V_m) is constant.

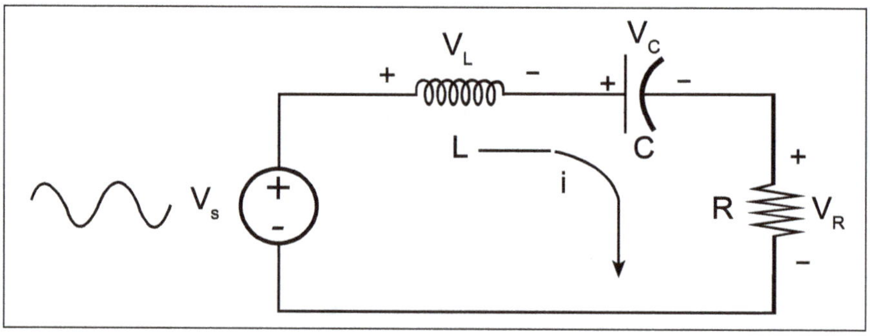

Series resonant circuit.

From the above figure, we have:

$$V_S = V_m \cos(\omega t)$$

$$i = I_m \cos(\omega t + \theta)$$

Frequency at which the reactance of the inductance and the capacitance cancel each other is the resonant frequency of this circuit.

This occurs at $\omega_o = 1/\sqrt{(LC)}$

Since $i = V_R / R$, the current i can be studied by studying the voltage across the resistor.

Current i is given as:

$$i = I_m \cos(\omega t + \theta)$$

$$I_m = \frac{V_m}{\sqrt{R^2 + \left(\omega L - \frac{1}{\omega C}\right)^2}}$$

$$\theta = -\tan^{-1}\left(\frac{\omega L - \frac{1}{\omega C}}{R}\right)$$

Bandwidth is given as:

$$B = \omega_2 - \omega_1 = R / L$$

$$\omega_{2,1} = \sqrt{\left(\frac{R}{2L}\right)^2 + \frac{1}{LC}} \pm \frac{R}{2L}$$

$\omega_{2,1}$ is the half power frequency or the 3 dB frequency (i.e.,) the frequencies at which I_m is equal to the maximum possible value divided by 1.414.

Quality factor is given as:

$$Q = \frac{\omega_0}{B} = \frac{1}{R}\sqrt{\frac{L}{C}}$$

Maximum value of V_L occurs at $\omega = \frac{\omega_0}{\sqrt{1 - \frac{R^2 C}{2L}}}$

V_C occurs at $\omega_0 \sqrt{1 - \frac{R^2 C}{2L}}$

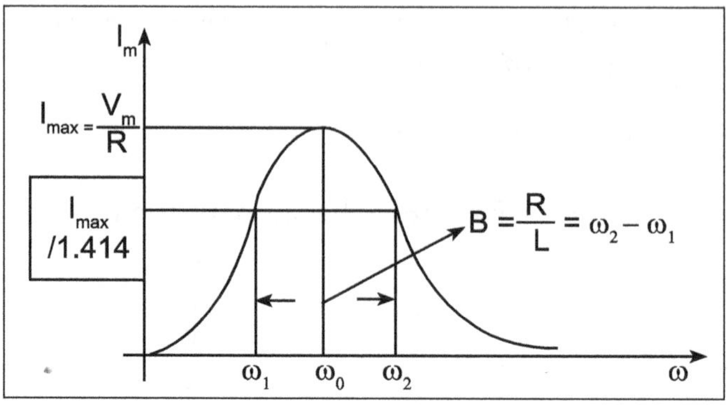

Frequency response of a series - resonant circuit.

Parallel Resonance

The basic parallel-resonant circuit is shown in the below figure. The steady state amplitude and the phase angle of the output voltage (V_o) vary with the frequency of the sinusoidal voltage source.

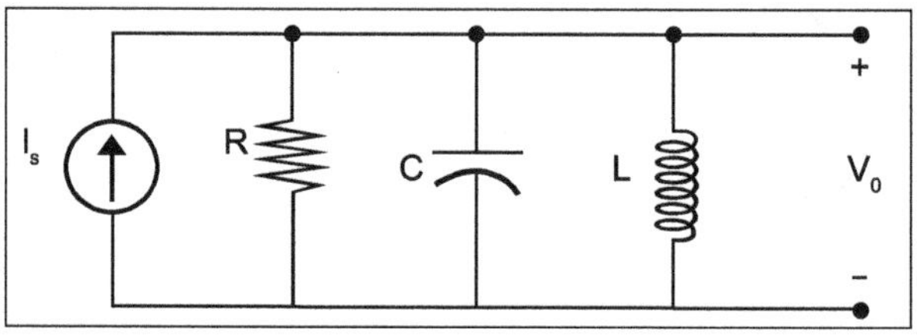

Parallel resonant circuit

From the above figure, we have:

If $I_s = I_m \cos(\omega t)$, then,

$$V_o = V_m \cos(\omega t + \theta)$$

Where,

$$V_m = \frac{I_m}{\sqrt{\frac{1}{R^2} + \left(\omega C - \frac{1}{\omega L}\right)^2}}$$

$$\theta = -\tan^{-1}\left(R\left(\omega C - \frac{1}{\omega L}\right)\right)$$

Resonant frequency is given as:

$$\omega_o = \frac{1}{\sqrt{LC}}$$

3 dB frequency is given as:

$$\omega_{2,1} = \sqrt{\left(\frac{1}{2RC}\right)^2 + \frac{1}{LC}} \pm \frac{1}{2RC}$$

Bandwidth is given as:

$$B = \omega_2 - \omega_1 = 1/RC$$

Quality factor:

$$Q = \omega_o / B = R\sqrt{(C/L)}$$

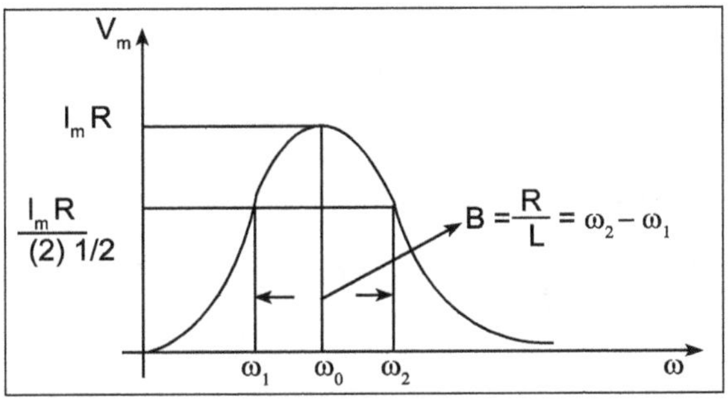

Frequency response of parallel resonant circuit.

Terms affected by frequency	R Independent of frequency		$jX_L = j_{\omega L}$ Directly proportional to frequency		$-jX_C = -j/\omega C$ Inversely proportional to frequen- cy	
Frequency	Magni- tude $\lvert M\rvert = R$	Phase	Magnitude $\lvert M\rvert = \omega L$	Phase $= \tan^{-1}(I_m/R)$	Magnitude $\lvert M\rvert = 1/\omega C$	Phase $= \tan^{-1}(I_m/R)$
$\omega = 0$	R	$0°$	$\lvert M\rvert = 0$	$\phi = \tan^{-1}(\omega_L/0)$ $\phi = \tan^{-1}(\infty)$ $\phi = 90°$	$\lvert M\rvert = -\infty$	$\phi = -\tan^{-1}(\omega_C/0)$ $\phi = -\tan^{-1}(\infty)$ $\phi = -90°$
$\omega = \infty$	R	$0°$	$\lvert M\rvert = \infty$		$\lvert M\rvert = -0$	

3.2 Variation of Impedance with Frequency

The impedance of series RLC circuit is given as:

$$Z = R + j\left(\omega L - \frac{1}{\omega C}\right)$$

Magnitude of impedance is given as:

$$|Z| = \sqrt{R^2 + \left(\omega L - \frac{1}{\omega C}\right)^2} \quad \text{...(1)}$$

At resonant frequency (f_0), the magnitude of impedance |Z| is equal to R where the reactive part is zero. The impedance of series RLC circuit is minimum at resonance and it is equal to R.

The variation of magnitude of impedance with frequency is shown in the below figure (b). The impedance under resonance is minimum which is denoted as Z_0.

It is given as:

$$Z_0 = R \quad \text{... (2)}$$

Current in series resonant circuit is given as:

$$I = \frac{V}{|Z|} = \frac{V}{\sqrt{R^2 + \left(\omega L + \frac{1}{\omega C}\right)^2}} \quad \text{...(3)}$$

Since the impedance is minimum at resonant frequency, current flowing in the circuit is maximum and it is denoted as I_0.

The maximum current at resonance is given as:

$$\therefore I_0 = \frac{V}{Z_0} = \frac{V}{R} \quad \text{...(4)}$$

The variation of current with frequency is shown in the below figure (a). The value of current is limited by the circuit resistance. If resistance is very small, then the current at resonance rises sharply to a very high value.

Since the admittance is the reciprocal of the impedance, the variation of admittance |Y| is opposite to that of variation of impedance of series RLC circuit. The variation of admittance |Y| is shown in the below figure(c).

At resonant frequency, admittance |Y| is maximum. At frequencies below resonant frequency, the current leads voltage because capacitive reactance is greater than inductive reactance.

So, phase angle is negative. This phase angle decreases as the frequency approaches resonant frequency. At resonance, phase angle is equal to 0°. At frequencies greater than resonant frequency, voltage leads current where inductive reactance is greater than the capacitive reactance.

So, phase angle is positive. As the frequency increases to very higher value, the phase angle approaches 90°. The phase angle variation with frequency is shown in the below figure (d).

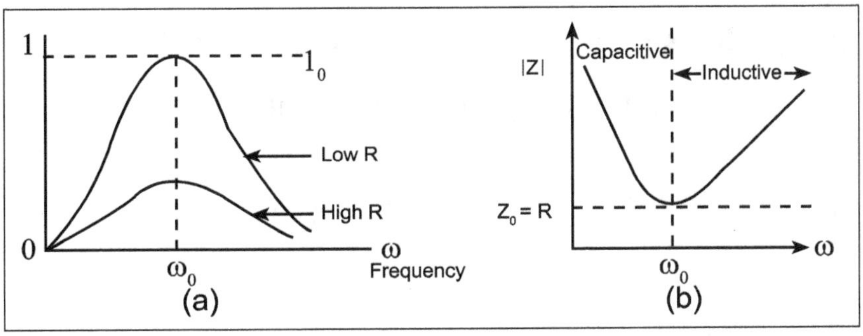

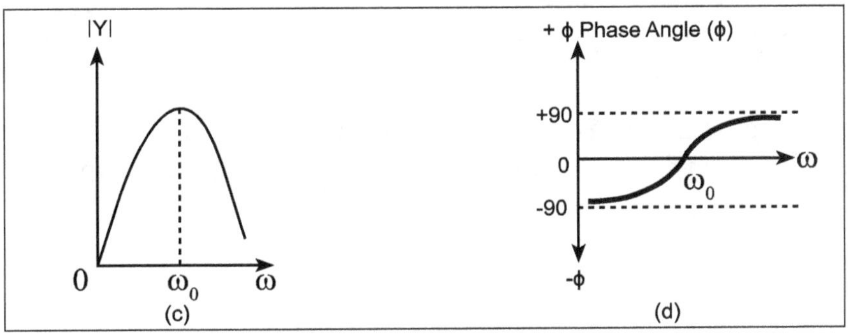

Variation of impedance, admittance and current with frequency.

Voltages Across L and C at Resonance

At resonance, the current through the series circuit is $I_o = \frac{V}{R}$. The magnitude of voltage across the inductance is given as:

$$V_L = I_o X_L = I_o(\omega_o L) = \frac{V}{R}\omega_o L = \left(\frac{\omega_o L}{R}\right)V$$

$$V_L = Q_o.V$$

Magnitude of voltage across the capacitance is given as:

$$V_C = I_o X_C = \frac{V}{R}\left(\frac{1}{\omega_o C}\right) = \left(\frac{1}{\omega_o RC}\right)V$$

$$V_C = Q_o.V$$

Magnitude of voltages that appears across L and C is equal to the multiplication of applied voltage by Q_o which is a number that is greater than unity. The voltages developed across L and C are more than the applied voltage only at resonance.

These voltages are equal in magnitude but opposite in phase and cancels each other at resonance. At resonance, series RLC circuit acts as a voltage amplifier. Hence, Q_o is also referred as the magnification factor of the circuit.

Frequencies for Maximum Voltage Across L and C

Let us consider that voltage across capacitor is V_C and it is given as:

$$V_C = I\left(\frac{I}{\omega C}\right), \quad \text{but } I = \frac{V}{Z}$$

$$\therefore V_C = \frac{V}{\omega C\sqrt{R^2 + \left(\omega L - \frac{1}{\omega C}\right)^2}} \qquad \text{...(1)}$$

To find frequency at which V_C is maximum, we have to differentiate V_C with respect to ω and equate it to zero where radical sign is removed by squaring expression.

$$\frac{dV_C^2}{d\omega} = 0$$

When V_C^2 is maximum, V_c is maximum.

By squaring the above equation (1), we have:

$$V_C^2 = \frac{V^2}{\omega^2 C^2\left[R^2 + \left(\omega L - \frac{1}{\omega C}\right)^2\right]}$$

$$\therefore V_C^2 = \frac{V^2}{\omega^2 C^2\left[R^2 + \left(\frac{\omega^2 LC - 1}{\omega^2 C^2}\right)^2\right]}$$

$$\therefore V_C^2 = \frac{V^2}{\omega^2 R^2 C^2 + \left(\omega^2 LC - 1\right)^2}$$

Now, differentiating V_C^2 with respect to ω and equating it to zero, we have:

$$\frac{dV_C^2}{d\omega} = \frac{V^2\left[2\omega R^2 C^2 + 2\left(\omega^2 LC - 1\right)\left(2\omega LC\right)\right]}{\left[\omega^2 R^2 C^2 + \left(\omega^2 LC - 1\right)^2\right]^2} = 0$$

Equating numerator terms to zero, we have:

$$V^2\left[2\omega R^2C^2+2\left(\omega^2LC-1\right)\left(2\omega LC\right)\right]=0$$

V is input voltage which cannot be zero.

$$\therefore 2\omega R^2C^2+2\left(2\omega LC\right)\left(\omega^2LC-1\right)=0$$

$$2\omega R^2C^2+4\omega^3L^2C^2-4\omega LC=0$$

$$4\omega^3L^2C^2=4\omega LC-2\omega R^2C^2$$

$$\therefore\ \omega^2=\frac{4\omega LC}{4\omega L^2C^2}-\frac{2\omega R^2C^2}{4\omega L^2C^2}$$

$$\therefore \omega^2=\frac{1}{LC}-\frac{R^2}{2L^2}$$

$$\therefore \omega=\sqrt{\frac{1}{LC}-\frac{R^2}{2L^2}}\ \text{rad/sec}$$

Therefore, the frequency f_C at which capacitor voltage v_C is maximum is given as:

$$f_C=\frac{1}{2\pi}\sqrt{\frac{1}{LC}-\frac{R^2}{2L^2}}\ \text{Hz} \qquad ...(2)$$

Simplifying the above equation (2), we get:

$$f_C=\frac{1}{2\pi\sqrt{LC}}\sqrt{1-\frac{R^2C}{2L}}$$

$$\therefore f_C=f_0\cdot\sqrt{1-\frac{R^2C}{2L}}\ \text{Hz} \qquad ...(3)$$

Let us calculate the frequency at which the voltage across the inductance is at its maximum.

Voltage across inductor (V_L) is given as:

$$V_L=I.(\omega L)\qquad \text{but } I=\frac{V}{Z}$$

$$\therefore\ V_L=\frac{V(\omega L)}{\sqrt{R^2+\left(\omega L-\frac{1}{\omega C}\right)^2}} \qquad ...(4)$$

Squaring the above equation (4), we have:

$$V_L^2 = \frac{V^2 \omega^2 L^2}{R^2 + \left(\omega L - \frac{1}{\omega C}\right)^2}$$

$$\therefore V_L^2 = \frac{V^2 \omega^2 L^2}{\frac{R^2 C^2 \omega^2 + \left(\omega^2 LC - 1\right)^2}{\omega^2 C^2}}$$

$$\therefore V_L^2 = \frac{V^2 \cdot \omega^4 L^2 C^2}{\omega^2 R^2 C^2 + \left(\omega^2 LC - 1\right)^2}$$

By differentiating V_L^2 with respect to ω and equating numerator term to zero, we have:

$$2 w^2 LC - w^2 R^2 C^2 - 2 = 0$$

$$\therefore \omega^2 \left(2LC - R^2 C^2\right) = 0$$

$$\therefore \omega^2 = \frac{2}{2LC - R^2 C^2}$$

$$\therefore \omega^2 = \frac{1}{LC - \frac{R^2 C^2}{2}}$$

$$\therefore \omega = \frac{1}{\sqrt{LC - \frac{R^2 C^2}{2}}} \text{rad/sec}$$

Therefore, the frequency f_L at which inductor voltage V_L is maximum is given as:

$$f_L = \frac{1}{2\pi\sqrt{LC - \frac{R^2 C^2}{2}}} \text{Hz} \qquad ...(5)$$

Simplifying the above equation (5), we get:

$$f_L = \frac{1}{2\pi\sqrt{LC} \cdot \sqrt{1 - \frac{R^2 C}{2L}}}$$

$$\therefore f_L = \frac{f_o}{\sqrt{1-\frac{R^2 C}{2L}}} \text{Hz} \qquad ...(6)$$

The above equations (3) and (6) gives frequencies f_C and f_L at which inductor voltage and capacitor voltages are maximum in terms of resonant frequency f_o.

Problems

1. A coil of inductance 0.1 H and resistance of 10 Ω is connected in series with a capacitor of 0.1 μF. Let us find the frequency of resonance of the circuit. Let us also find the quality factor of the circuit at resonance.

Solution:

Given:

$L = 0.1$ H

$R = 10$ Ω

$C = 0.1$ μF

(i) Frequency of resonance is given as:

$$f_o = \frac{1}{2\pi\sqrt{LC}} = \frac{1}{2\times\pi\times\sqrt{(0.1)(0.1\times10^{-6})}}$$

$$\therefore f_o = 1.5915\times10^3 \text{ Hz} = 1.5915 \text{ kHz}$$

(ii) Quality factor of the circuit is given as:

$$Q_o = \frac{\omega_o L}{R} = \frac{(2\pi f_o)L}{R} = \frac{(2\times\pi\times1.5915\times10^3)(0.1)}{10}$$

$$\therefore Q_o = 99.99 \approx 100$$

$$Q_o = \frac{1}{R}\sqrt{\frac{L}{C}} = \frac{1}{10}\sqrt{\frac{0.1}{0.1\times10^{-6}}} = 100$$

2. A series RLC circuit consists of R = 100Ω, L = 0.02H and C = 0.02μF. Let us calculate frequency of resonance. A variable frequency sinusoidal voltage of 50 V is applied to the circuit. Let us also find the frequency at which voltage across L and C is maximum. Let us also calculate the voltage across L and C at frequency of resonance. Finally, let us find the maximum current in the circuit.

Solution:

Given:

R = 100Ω

L = 0.02 H

C = 0.02 μF

(i) Frequency of resonance is given as:

$$f_o = \frac{1}{2\pi\sqrt{LC}} = \frac{1}{2\times\pi\times\sqrt{(0.02)(0.02\times10^{-6})}}$$

$$\therefore f_o = 7.957\,kHz$$

(ii) The frequency (f_C) at which voltage across C is maximum is given as:

$$f_C = \frac{1}{2\pi}\sqrt{\frac{1}{LC} - \frac{R^2}{2L^2}}$$

$$f_C = \frac{1}{2\pi}\sqrt{\frac{1}{(0.02)(0.02\times10^{-6})} - \frac{(100)^2}{2(0.02)^2}}$$

$$\therefore f_C = 7.937\,kHz$$

Thus, the maximum voltage that appears across capacitance at a frequency f_C is less than the resonant frequency f_o.

(iii) The frequency f_L at which voltage across L is maximum is given as:

$$f_L = \frac{1}{2\pi\sqrt{LC - \frac{R^2C^2}{2}}}$$

$$\therefore f_L = \frac{1}{2\pi\sqrt{(0.02)(0.02\times10^{-6}) - \frac{(100)^2(0.02\times10^{-6})^2}{2}}}$$

$$\therefore f_L = 7.977\,kHz$$

Thus, the maximum voltage appears across inductance at a frequency f_L which is less than the resonant frequency f_o.

(iv) Voltage across L and C at resonance is given as:

$$V_C = -jQ_o V$$

$$V_L = -jQ_o V$$

Quality factor of the circuit (Q_o) is given as:

$$Q_o = \frac{\omega_o L}{R} = \frac{(2\times\pi\times f_o)L}{R} = \frac{(2\times\pi\times 7.957\times 10^3)(0.02)}{100}$$

$$\therefore Q_o = 9.999 \approx 10$$

Voltage across capacitor is given as:

$$V_C = -jQ_o V = -j(10)(50) = -j500\,V = 500\angle -90°\,V$$

Voltage across inductor is given as:

$$V_L = jQ_o V = j(10)\,(50) = j500\,V = 500\,\angle -90°\,V$$

(v) Maximum current in the circuit is given as:

$$I_o = \frac{V}{R} = \frac{50}{100} = 0.5\,A$$

3.3 Bandwidth, Q Factor and Selectivity

Bandwidth

It is defined as a band that contains all frequencies between upper cut-off and lower cut-off frequencies.

At resonance, the voltages across the capacitor and the pure inductance cancel out where the series impedance takes its minimum value (Z_o = R). If the voltage is kept constant, then the current is maximum at resonance. The current goes to zero at low frequency because X_C is infinite. The current also goes to zero at high frequency since X_L increases with ω.

The bandwidth (B) is defined as the difference between the two frequencies ω_1 and ω_2 at which the circuit converts power at half the maximum rate.

Bandwidth is given as:

$$B = \omega_2 - \omega_1$$

At a certain frequency, the power dissipated by the resistor is half of the maximum power which occurs at $\omega_o = 1/\sqrt{(LC)}$.

The half power occurs at the frequencies for which the amplitude of the voltage across the resistor is equal to $1/\sqrt{2}$ of the maximum value.

$$P_{1/2} = \frac{1}{4}\frac{V_{max}^2}{R}$$

The below figure shows the graphical form of various frequencies:

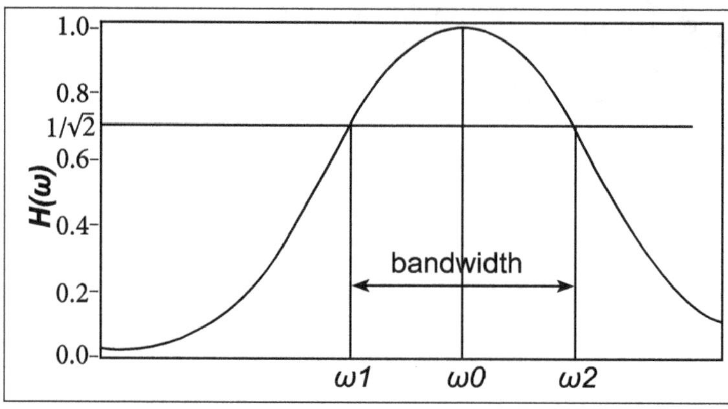

Graphical representation of various frequencies.

Half power occurs at the following condition.

$$\frac{1}{\sqrt{2}} = \frac{\omega RC}{\sqrt{(1-\omega^2 LC)^2 + (\omega RC)^2}}$$

Equation has two roots namely:

$$\omega_1 = -\frac{R}{2L} + \sqrt{\left(\frac{R}{2L}\right)^2 + \frac{1}{\omega_o^2}}$$

$$\omega_2 = \frac{R}{2L} + \sqrt{\left(\frac{R}{2L}\right)^2 + \frac{1}{\omega_o^2}}$$

Bandwidth is given as:

$$B = \omega_2 - \omega_1$$

By multiplying these two equations, we can show that ω_o is the geometric mean of ω_1 and ω_2.

$$\omega_o = \sqrt{(\omega_1\omega_2)}$$

Bandwidth increases with increasing R and the sharpness of the resonance increases with decreasing R.

For a fixed L and C, decrease in R corresponds to a narrower resonance and a higher selectivity regarding the frequency range is passed by the circuit.

If we increase R, the frequency range over which the dissipative characteristics dominate the behavior of the circuit increases. In order to quantify this behavior, we define a parameter called quality factor (Q) which is related to the sharpness of the peak and it is given as:

$$Q = 2\pi \frac{\text{maximum energy stored}}{\text{total energy lost per cycle at resonance}} = 2\pi \frac{E_s}{E_n}$$

The above equation represents the ratio of the energy stored to the energy dissipated in a circuit.

The energy stored in the circuit is given as:

$$E_s = \frac{1}{2}LI^2 + \frac{1}{2}CV_C^2$$

Since $V_c = A\sin\omega t$, the current flowing in the circuit is expressed as:

$$I = C\frac{dV_C}{dt} = \omega CA\cos(\omega t)$$

Total energy stored in the reactive element is given as:

$$E_S = \frac{1}{2}L\omega^2 C^2 A^2 \cos^2(\omega t) + \frac{1}{2}CA^2 \sin^2(\omega t)$$

At the resonance frequency:

$$\omega = \omega_o = \frac{1}{\sqrt{LC}}$$

Energy stored in the circuit is given as:

$$E_S = \frac{1}{2}CA^2$$

Energy dissipated per period is equal to the average resistive power dissipated times the oscillation period.

$$E_D = R\langle I^2\rangle \frac{2\pi}{\omega_0} = R\left(\frac{\omega_0^2 C^2 A^2}{2}\right)\frac{2\pi}{\omega_0} = 2\pi\left(\frac{1}{2}\frac{RC}{\omega_0 L}A^2\right)$$

$$Q = \omega_0 L / R$$

By combining these equations, the relationship between the bandwidth and the Q factor is obtained.

Comparison of RLC Resonant Circuits

Categories	**Series**	**Parallel**
Circuit	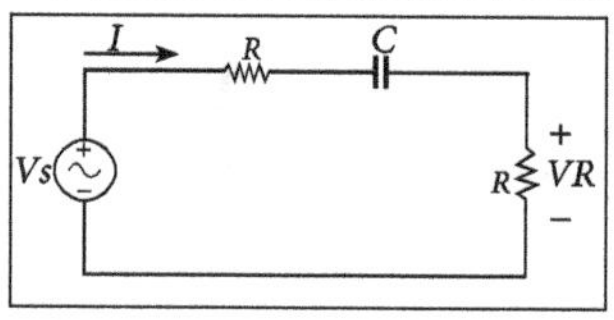	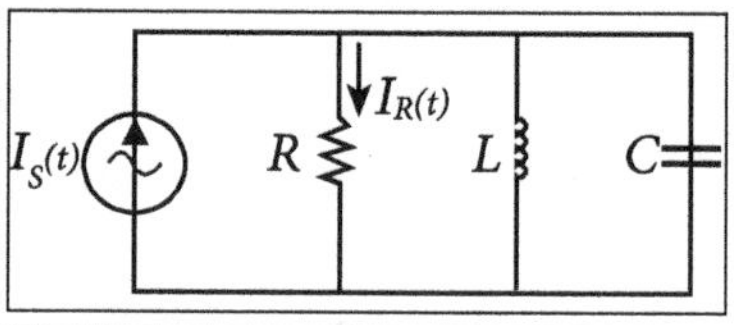
Transfer function	$\lvert H(\omega)\rvert \equiv \left\lvert\frac{VR}{V_S}\right\rvert = \frac{\omega RC}{\sqrt{(1-\omega^2 LC)^2 + (\omega RC)^2}}$	$\lvert H(\omega)\rvert \equiv \left\lvert\frac{I_R}{I_S}\right\rvert = \frac{\omega L}{\sqrt{(R-\omega^2 LCR)^2 + (\omega L)^2}}$
Resonant frequency	$\omega_0 = \frac{1}{\sqrt{LC}}$	$\omega_0 = \frac{1}{\sqrt{LC}}$
1/2 power frequencies	$\omega_1 = -\frac{R}{2L} + \sqrt{\left(\frac{R}{2L}\right)^2 + \frac{1}{\omega_0^2}}$ $\omega_2 = -\frac{R}{2L} + \sqrt{\left(\frac{R}{2L}\right)^2 + \frac{1}{\omega_0^2}}$	$\omega_1 = -\frac{1}{2RC} + \sqrt{\left(\frac{1}{2RC}\right)^2 + \frac{1}{\omega_0^2}}$ $\omega_2 = -\frac{1}{2RC} + \sqrt{\left(\frac{1}{2RC}\right)^2 + \frac{1}{\omega_0^2}}$
Bandwidth	$B_S = \omega_2 - \omega_1 = \frac{R}{L}$	$B_P = \omega_2 - \omega_1 = \frac{1}{RC}$
Q factor	$Q = \frac{\omega_0}{B_S} = \frac{\omega_0 L}{R} = \frac{1}{\omega_0 RC}$	$Q = \frac{\omega_0}{B_P} = \omega_0 RC = \frac{R}{\omega_0 L}$

Quality Factor (Q)

Reactive components such as capacitors and inductors are often described with a figure of merit called Q.

$$Q = \omega \frac{\text{energy stored}}{\text{average power dissipated}}$$

Quality factor (Q) is the ratio of the reactive power in the inductor or capacitor to the true power in the resistance in series with the coil or capacitor.

$$Q = 2\pi \times \frac{\text{max energy stored}}{\text{energy dissipated per cycle}}$$

Quality factor of the capacitance circuit $Q = \frac{1}{wCR}$

Thus, it is a measure of the ratio of stored energy to the lost energy per unit time. An ideal reactive component (capacitor or inductor) stores energy which is given as:

$$E = \frac{1}{2} CV_{pk}^2 \text{ or } \frac{1}{2} LI_{pk}^2$$

Since any real component has loss due to the resistive component, the average power dissipated is given as:

$$P_{avg} = \frac{1}{2} I_{pk}^2 R = \frac{V_{pk}^2}{2R}$$

A series resonant circuit is shown in the below figure.

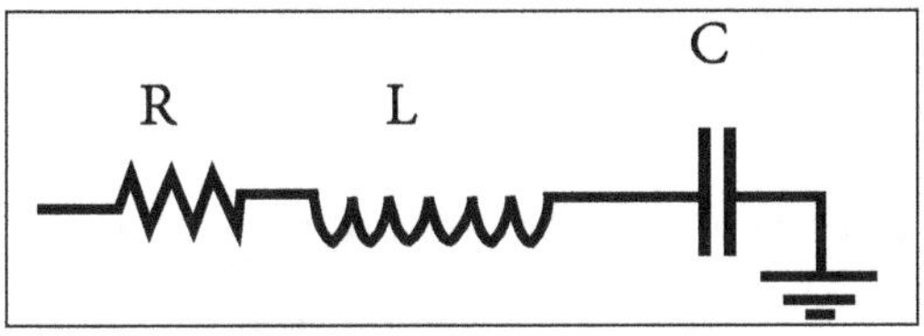

Series resonant circuit.

At resonance, the reactances cancel out which leaves a peak voltage (V_{pk}) across the loss resistance(R).

Thus,

$$I_{pk} = V_{pk} / R$$

I_{pk} is the maximum current which passes through all elements.

Then,

$$Q = \omega_o \frac{LI_{pk}^2 / 2}{I_{pk}^2 R / 2} = \frac{\omega_o L}{R} = \frac{1}{\omega_o RC}$$

$$Q = \frac{1}{\omega RC} = \frac{X}{R}$$

Capacitor is resonated with an ideal inductor at frequency ω. X is the capacitive reactance and R is the series resistance. Since this Q refers only to the capacitor itself in isolation from the rest of the circuit, it is called as unloaded Q or Q_U. The higher the unloaded Q, the lower the loss where Q decreases with frequency.

Unloaded Q of an inductor is given as:

$$Q_U = \frac{\omega_0 L}{R}$$

Where,

R = Series resistance

Q is proportional to frequency of an inductor. Q of an inductor will depend on the wire diameter, core material (air, powdered iron, ferrite) and checks whether inductor is in a shielded metal can or not.

For a parallel resonant circuit, Q is the ratio of susceptance to conductance.

$$Q = B/G$$

Where,

B = Susceptance of the capacitor or inductor

G = Shunt conductance

Quality Factor for Series and Parallel RLC Circuit

Circuit	Bandwidth	Capacitor	Inductor
Quality factor of series RLC circuit	$Q=\frac{f_r}{f_2-f_1}=\frac{f_r}{BW}=\frac{f_r}{R/2\pi L}$	$Q=\frac{1}{wCR}$	$Q=\frac{wL}{R}$
Quality factor of parallel RLC circuit	$Q=\frac{f_r}{f_2-f_1}=\frac{f_r}{BW}=\frac{f_r}{1/RC}=f_r RC$	$Q=wCR$	$Q=\frac{R}{wL}$

Selectivity

The ratio of bandwidth to the resonant frequency is called as selectivity.

$$S=\frac{B.W}{f_r}$$

3.4 Inductance

Self-inductance

Self-inductance of a coil is defined as the ratio of self-induced emf to the rate of change of current in the coil. It is denoted by 'L' and it depends on the physical characteristics of the coil. Unit of self-inductance is Henry.

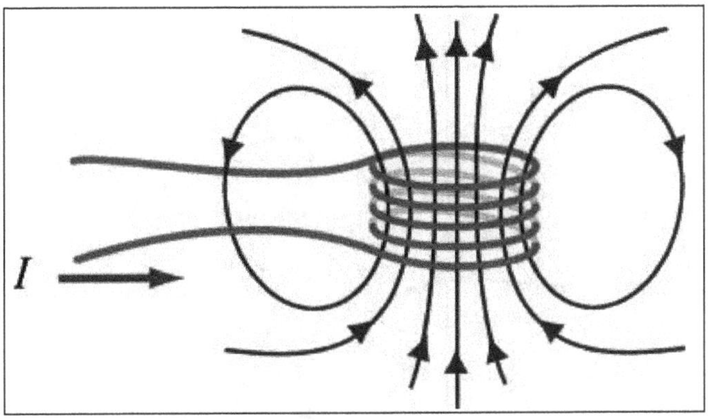

Magnetic flux through the current loop.

Let us consider a coil consists of N turns and carrying current I in the counterclockwise direction as shown in the below figure. If the current is steady, then the magnetic flux through the loop will remain constant. Suppose the current I changes with time, then according to Faraday's law, an induced emf will arise to oppose the change.

The induced current will flow clockwise if $dI/dt > 0$ and counterclockwise if $dI/dt < 0$. The property of the loop in which its own magnetic field opposes any change in current is termed as self-inductance and the emf generated is called as the self-induced emf or back emf. All current-carrying loops exhibit this property. In particular, an inductor is a circuit element which has a large self-inductance.

Self-induced emf is written as:

$$\varepsilon_L = -N\frac{d\Phi_B}{dt} = -N\frac{d}{dt}\iint \vec{B}\cdot d\vec{A}$$

$$\varepsilon_L = -L\frac{dI}{dt}$$

$$L = \frac{N\Phi_B}{I}$$

Inductance (L) is a measure of an inductor's resistance to the change of current. Larger the value of L, lower is the rate of change of current.

Mutual Inductance

It is a property associated with two or more coils or inductors which are in close presence of common magnetic flux which links the coils. A transformer is such a device whose operation is based on mutual inductance.

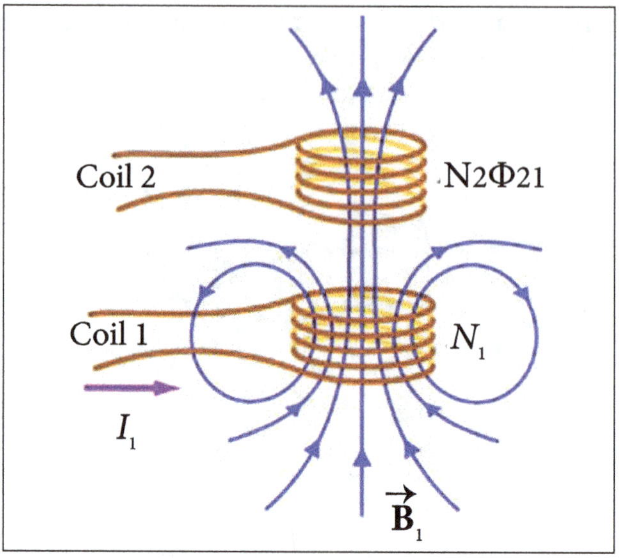

Changing current in coil 1 produces changing magnetic flux in coil 2.

Let us consider two coils L_1 & L_2 which are close together where the flux produced by i_1 in coil L_1 links coil L_2. When a voltage V_1 is applied across L_1, a current I_1 will start flowing through L_1 which produces a flux ϕ. This flux ϕ links coil L_2. If the current flowing through L_1 changes with respect to time, then the induced flux ϕ also changes with respect to time.

The time varying flux surrounding the second coil L_2 induces an electromagnetic force (emf) or voltage across the coil L_2. This voltage is proportional to the rate of change of current flowing through coil L_1. Then, the two coils are said to be inductively coupled and it is called as mutual inductance.

Let the two coils are placed near to each other as shown in the above figure. The first coil has N_1 turns and carries a current I_1 which gives rise to a magnetic field B1. Since the two coils are close to each other, some of the magnetic field lines through coil 1 will also pass through coil 2.

Let ϕ_{21} denote the magnetic flux through one turn of coil 2 due to I_1. By varying I_1 with time, there will be an induced emf associated with the changing magnetic flux in the second coil.

$$\varepsilon_{21} = -N_2 \frac{d\Phi_{21}}{dt} = -\frac{d}{dt} \iint_{coil\,2} \vec{B}_1 \cdot d\vec{A}_2$$

The time rate of change of magnetic flux ϕ_{21} in coil 2 is proportional to the time rate of change of the current in coil 1.

$$N_2 \frac{d\Phi_{21}}{dt} = M_{21} \frac{dI_1}{dt}$$

Proportionality constant (M_{21}) = Mutual inductance

$$M_{21} = \frac{N_2 \Phi_{21}}{I_1}$$

SI unit for inductance is henry (H).

1 henry = 1 H = 1 T m²/A

The mutual inductance M_{21} depends only on the geometrical properties of the two coils such as the number of turns and the radii of the two coils. If the current (I_2) in the second coil varies with time, then the induced emf in coil 1 is given as:

$$\varepsilon_{12} = -N_1 \frac{d\Phi_{12}}{dt} = -\frac{d}{dt} \iint_{coil1} \vec{B}_2 \cdot d\vec{A}_1$$

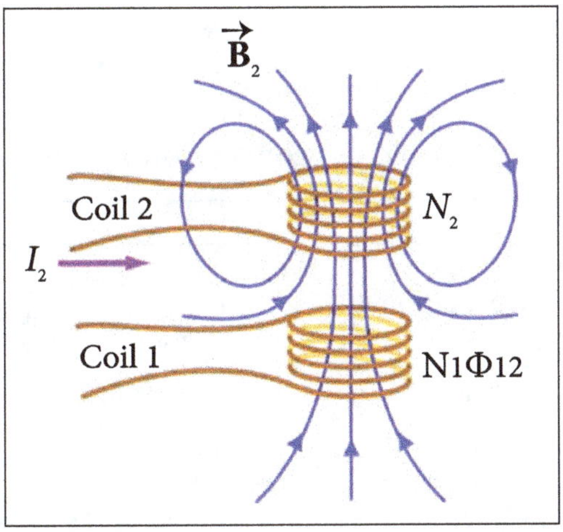

Changing current in coil 2 produces changing magnetic flux in coil 1.

This changing flux in coil 1 is proportional to the changing current in coil 2.

$$N_1 \frac{d\Phi_{12}}{dt} = M_{12} \frac{dI_2}{dt}$$

Proportionality constant M_{12} is another mutual inductance which is written as:

$$M_{12} = \frac{N_1 \Phi_{12}}{I_2}$$

Reciprocity theorem combines the Ampere's law and the Biot-Savart law where the constants are equal.

$$M_{12} = M_{21} = M$$

Dot Convention

In circuit analysis, the dot convention is the convention which is used to denote the voltage polarity of mutual inductance of two component.

The two conventions are given as:

- If a current enters the dotted terminal of a coil, the reference polarity of the voltage induced in the other coil is positive at its dotted terminal.
- If a current leaves the dotted terminal of a coil, the reference polarity of the voltage induced in the other coil is negative at its dotted terminal.

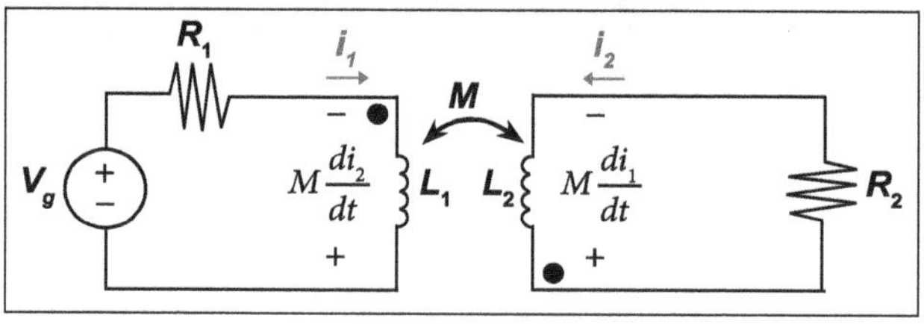

Dot convention.

The equivalent circuits of all possible dot convention are shown in the below figure.

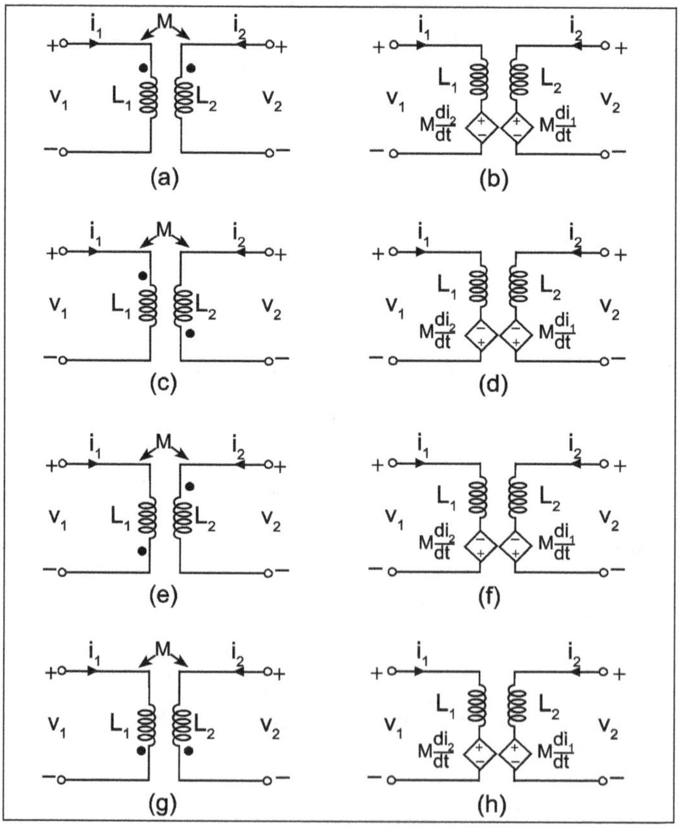

Equivalent circuits of all possible dot convention.

Magnetically Coupled Circuits and Equivalent Circuit with Different Dot Conventions

Let us consider a magnetically coupled circuit with dots placed as shown in the above figure (a). Both the currents i1 and i2 are entering the dotted terminals. According to the dot convention, the mutually induced e.m.f. in both the coils has same polarity as self-induced e.m.f. in respective coil. The equivalent circuit is shown in the above figure (b).

By applying KVL, the network equations of the equivalent circuit is written as:

$$v_1 = L_1 \frac{di_1}{dt} + M \frac{di_2}{dt}$$

$$v_2 = L_2 \frac{di_2}{dt} + M \frac{di_1}{dt} \qquad ...(1)$$

Let us consider magnetically coupled circuit as shown in the above figure (c) where dot is placed at lower terminal of coil L_2. Hence, current i_1 enters through dotted terminal of L_1 while current i_2 leaves through dotted terminal of L_2.

According to dot convention, the polarity of mutually induced e.m.f. in L_1 due to i_2 in L_2 is opposite to that of self-induced e.m.f. in coil L. The polarity of mutually induced e.m.f. in coil L_2 due to the current i_1 in coil L_1 will be opposite to that of self-induced e.m.f. in coil L_2. The equivalent circuit is shown in the above figure (d).

By using KVL, the network equations are written as:

$$v_1 = L_1 \frac{di_1}{dt} - M \frac{di_2}{dt}$$

$$v_2 = L_2 \frac{di_2}{dt} - M \frac{di_1}{dt} \qquad ...(2)$$

By applying KVL, the network equations can be written as:

$$v_1 = L_1 \frac{di_1}{dt} - M \frac{di_2}{dt}$$

$$v_2 = L_2 \frac{di_2}{dt} - M \frac{di_1}{dt} \qquad ...(3)$$

For last possible combination, both the dots are placed at lower terminals of coils L_1 and L_2 and both the currents leave dot as shown in the above figure (g). The equivalent circuit is shown in the above figure (h).

By applying KVL, the network equations are written as:

$$v_1 = L_1 \frac{di_1}{dt} + M\frac{di_2}{dt}$$

$$v_2 = L_2 \frac{di_2}{dt} + M\frac{di_1}{dt} \qquad ...(4)$$

The analysis of multi-winding inductor networks is carried out for each pair of windings using same dot convention.

Coefficient of Coupling

In coupled coils, the coefficient of coupling is defined as the fraction of the total flux produced by linking one coil with another coil.

The amount of coupling between the inductively coupled coils is expressed in terms of the coefficient of coupling which is defined as:

$$K = \frac{\varphi_{12}}{\varphi_1} = \frac{\varphi_{21}}{\varphi_2}$$

At K = 1, maximum mutual inductance occurs in coupled circuits:

Where,

M - Mutual inductance between the two coils.

L_1 – Self-inductance of the first coil.

L_2 – Self-inductance of the second coil.

Problems

1. An iron ring of relative permeability 100 is wound uniformly with two coils of 100 and 400 turns of wire. The cross-section of the ring is 4cm². The mean circumference is 50cm.

Let us calculate the following:

1. Self-inductance of each of the two coils.
2. Mutual inductance.
3. Total inductance when the coils are connected in series with the flux in the same sense.
4. Total inductance when the coils are connected in series with flux in the opposite sense.

Solution:

Given:

Relative permeability = 100

Number of turns of wire = 100 and 400

Cross-section of the ring = 4 cm^2

Mean circumference = 50 cm

Formula to be used:

$$L=\frac{\mu_o N^2 A}{2\pi R_m}$$

$$M=\frac{\mu_o N_1 N_2 A}{2\pi R_m}$$

Total inductance = $L_1 + L_2 + 2M$

(1) Self-inductance of coil with 100 turns is given as:

$$L=\frac{\mu_o N^2 A}{2\pi R_m}=\frac{4\pi\times10^{-7}\times100^2\times4}{50\times10^{-2}}=0.1H$$

Self-inductance with 400 turns is given as:

$$L=\frac{4\pi\times10^{-7}\times400^2\times4}{50\times10^{-2}}=1.61\,\text{Henry}$$

(2) Mutual inductance of the coil is given as:

$$M=\frac{\mu_o N_1 N_2 A}{2\pi R_m}$$

$$=\frac{4\pi\times10^{-7}\times100\times400\times4}{50\times10^{-2}}$$

$$= 0.402\,H$$

(3) Total inductance = $L_1 + L_2 + 2M$

= 0.1 + 1.61+ 2x0.4

= 2.51

(4) Total inductance when the coils are in series with flux in the opposite sense is given as,

$$L_1 + L_2 - 2M = 0.1 + 1.61 - 2 \times 0.4$$

$$= 0.91.$$

2. A long cylindrical solenoid with 100 loops per 1 cm has a radius of 1.6 cm. Let us assume that the magnetic field inside the solenoid is homogeneous and parallel to the axis of the solenoid. Let us find the following quantities.

1. Inductance of the solenoid per 1 meter of its length.
2. Electromotive force induced per 1 meter of the length of the solenoid if the current change is 13 As^{-1}.

Solution:

1. Inductance of the solenoid is defined as the constant of proportionality between the total magnetic flux through the coil turns and the current flowing in the coil.

$$N\phi = LI$$

Where,

N = Number of turns per one meter (N = 10,000)

The magnetic flux φ is the flux through one turn. Since the magnetic field is homogeneous, we have,

$$\phi = BS\cos\alpha$$

The magnetic induction (B) at all points is perpendicular to the cross section of the coil.

$$\cos\alpha = 1$$

Where,

α = Angle between the magnetic induction B and normal to the surface.

Circular cross-sectional area of solenoid is given as:

$$S = \pi R^2$$

Magnetic induction inside the coil for the magnetic induction B is given as:

$$B = \mu_0 NI$$

$$N\mu_0 NI\pi R^2 = LI$$

Unknown inductance (L) is given as:

$$L = \mu_o \pi N^2 R_2$$

2. For the electromotive force induced in the coil, we have:

$$|U_i| = |d(N\phi)/dt|$$

For the inductance (L) of a one meter long coil, we have,

$$N\phi = LI$$

$$|U_i| = L(dI/dt)$$

3.4.1 Analysis of Multi Winding Coupled Circuits

Inductively coupled multi-mesh circuits are analysed by using Kirchhoff's laws and by loop current methods. Let us consider the below figure where three coils are inductively coupled. For such a system of inductors, inductance matrix (L) is given as:

$$L = \begin{bmatrix} L_{11} & L_{12} & L_{13} \\ L_{21} & L_{22} & L_{23} \\ L_{31} & L_{32} & L_{33} \end{bmatrix}$$

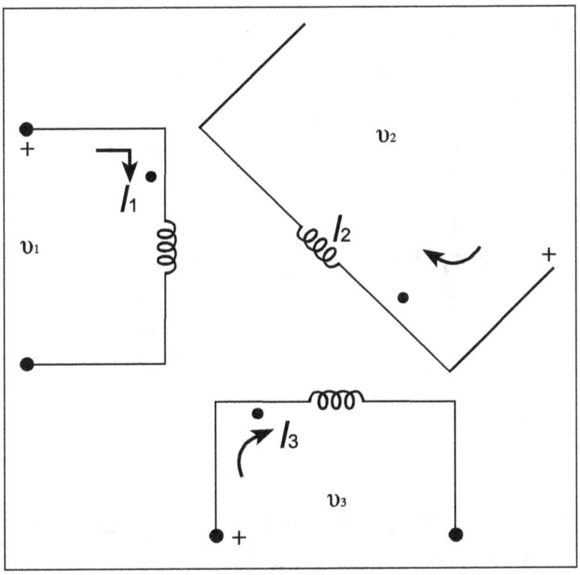

Inductively coupled coils.

L_{11}, L_{22} and L_{33} are self-inductances of the coupled circuits and $L_{12} = L_{21}$, $L_{23} = L_{32}$ and $L_{13} = L_{31}$ are mutual inductances.

L_{12} is the mutual inductance between coils 1 and 2, L_{13} is the mutual inductance between coils 1 and 3 and L_{23} is the mutual inductance between coils 2 and 3.

$$[v]=[L]\left[\frac{di}{dt}\right]$$

Where,

v and i = Vectors of the branch voltages and currents.

Branch volt-ampere relationships of the three inductors are given as:

$$\begin{bmatrix} v_1 \\ v_2 \\ v_3 \end{bmatrix} = \begin{bmatrix} L_{11} & L_{12} & L_{13} \\ L_{21} & L_{22} & L_{23} \\ L_{31} & L_{32} & L_{33} \end{bmatrix} \begin{bmatrix} di_1/dt \\ di_2/dt \\ di_3/dt \end{bmatrix}$$

Using KVL and KCL, the effective inductances can be calculated. The polarity for the inductances is determined by using passivity criteria, whereas, the signs of the mutual inductances can be determined by using the dot convention.

Problem

1. For the circuit shown in the below figure, let us write the inductance matrix.

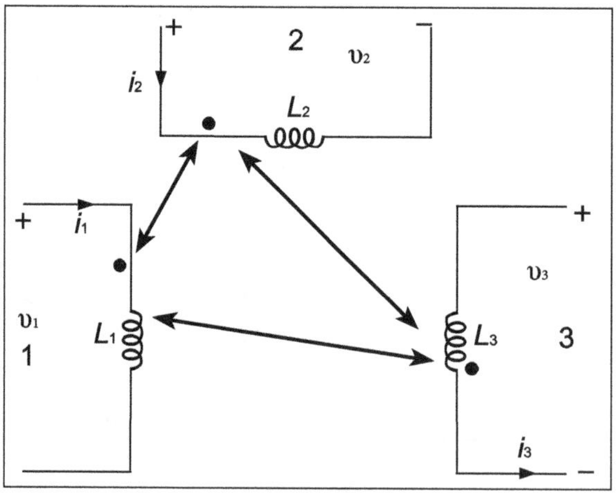

Solution:

Let L_1, L_2 and L_3 are the self-inductances and $L_{12} = L_{21}$, $L_{23} = L_{32}$ and $L_{13} = L_{31}$ are the mutual inductances between coils 1-2, 2-3 and 1-3.

If $L_{12} = L_{21}$ is positive, then both the currents enters at dot marked terminals, whereas, $L_{13} = L_{31}$ and $L_{23} = L_{32}$ are negative.

The inductance matrix is given as:

$$L = \begin{bmatrix} L_1 & L_{12} & -L_{13} \\ L_{21} & L_2 & -L_{23} \\ -L_{31} & -L_{32} & L_3 \end{bmatrix}$$

3.5 Coupled Inductors and Circuits

Series Combination of Inductors

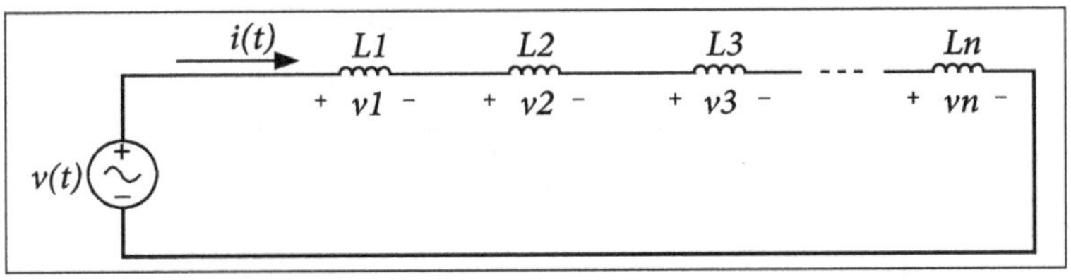

Series combination of inductors.

By applying KVL around the loop, we have:

$$v = v_1 + v_2 + v_3 + .. + v_n$$

$$= \Bigg[\underbrace{L_1 + L_2 + L_3 + ... + L_n}_{Leq} \Bigg] \frac{di}{dt}$$

$$= Leq \frac{di}{dt} \qquad ...(1)$$

Inductor in series combines like the resistor in series. The energy stored in an inductor is the integral of the instantaneous power delivered to the inductor. Let us assume that no current flows through the inductor at $t = -\infty [i(-\infty) = 0]$.

Energy stored in the inductor at time t is given as:

$$E(t) = \int_{-\infty}^{t} P(\tau) d\tau$$

$$= \int_{-\infty}^{t} v(\tau) i(\tau) d\tau$$

$$= \int_{-\infty}^{t} L \frac{di(\tau)}{d\tau} i(\tau) d\tau$$

$$=\frac{1}{2}Li(t)^2 \qquad ...(2)$$

Parallel Combination of Inductors

Inductors connected in series and in parallel are combined to form an equivalent inductance. Let us consider the parallel combination of inductors as shown in the below figure. All the inductors have the same voltage across them.

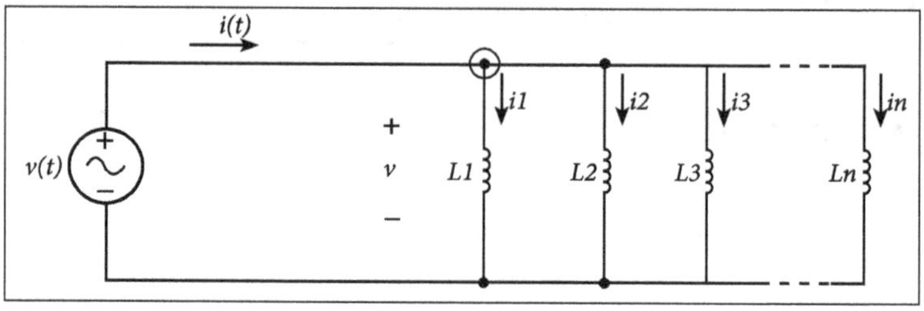

Parallel combination of inductors.

By applying KCL, we have:

$$i = i_1 + i_2 + i_3 + ... + i_n \qquad ...(3)$$

$$ik = \frac{1}{Lk}\int_0^t v\,dt + ik(0)$$

$$i = \frac{1}{L1}\int_0^t v\,dt + i1(0) + \frac{1}{L2}\int_0^t v\,dt + i2(0) + \frac{1}{L3}\int_0^t v\,dt + i3(0) + ... + \frac{1}{Ln}\int_0^t v\,dt + in(0)$$

$$= \left(\underbrace{\frac{1}{L1} + \frac{1}{L2} + \frac{1}{L3} + ... + \frac{1}{Ln}}_{\frac{1}{Leq}}\right)\int_0^t v\,dt + \underbrace{i1(0) + i2(0) + i3(0) + ... + in(0)}_{I(0)}$$

$$= \frac{1}{Leq}\int_0^t v\,dt + i(0) \qquad ...(4)$$

Inductors in parallel are combined like the resistors in parallel.

3.5.1 Single Tuned and Double Tuned Coupled Circuits

Single Tuned Amplifier

A common emitter amplifier is converted into a single tuned amplifier by including a parallel tuned circuit as shown in the below figure. The biasing components are not shown for simplicity.

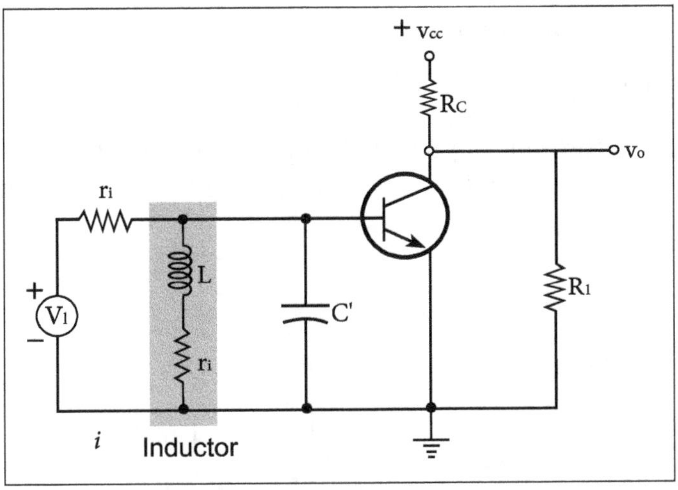

(a) Single tuned transistor amplifier.

Let us see the several practical assumptions to simplify the analysis.

Assumptions

A common emitter amplifier can be converted into a single tuned amplifier by including a parallel tuned circuit as shown in the below figure. The biasing components are not shown for simplicity.

Before going to study the analysis of this amplifier, let us see the several practical assumptions to simplify the analysis.

1. $R_L << R_C$

2. $r_{bb}' = 0$

Simplified equivalent circuit for a single tuned amplifier is shown in the below figure.

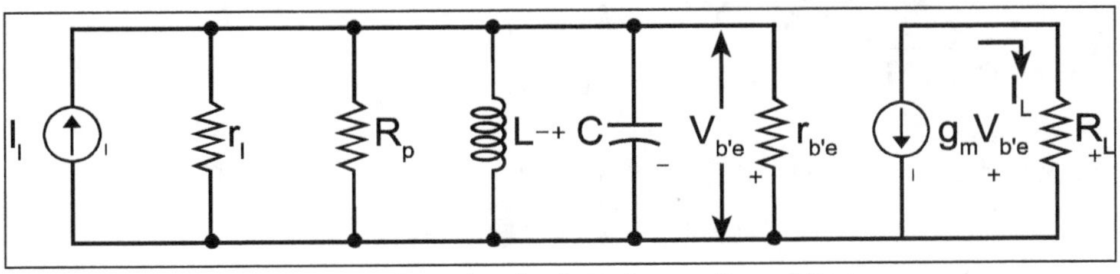

(b) Equivalent circuit of single tuned amplifier.

$$C_{eq} = C' + C_{b'e} + (1 + g_m R_L) C_{b'c}$$

Where,

C' - External capacitance used to tune the circuit

$(1+g_m R_1)\ C_{b'c}$ - Miller capacitance

r_s - Losses in coil

The series RL circuit in the above figure (a) is replaced by an equivalent circuit as shown in the below figure (b) where coil losses are low over the frequency band of the interest.

$$Q_C = \frac{\omega L}{r_c} >> 1 \qquad ...(1)$$

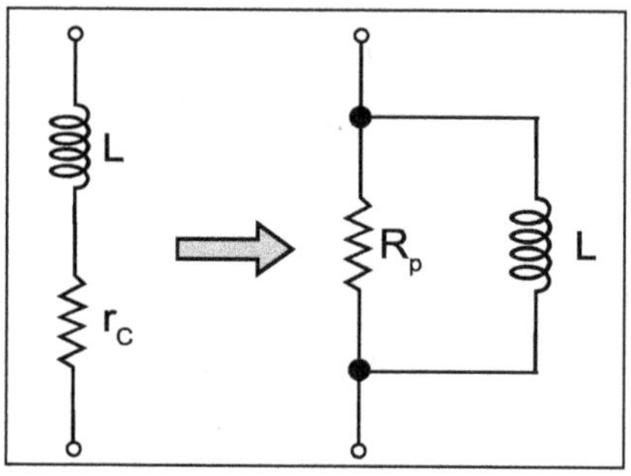

(c) Equivalent circuit.

The conditions for equivalence are most easily established by equating the admittances of the two circuits as shown in the above figure (c).

$$Y_1 = \frac{1}{r_c + j\omega L} = \frac{r_c - j\omega L}{r_c^2 + \omega^2 L^2}$$

$$\approx \frac{r_c}{r_c^2 + \omega^2 L^2} - \frac{j\omega L}{r_c^2 + \omega^2 L^2}$$

From the above equation (1):

$$\omega L >> r_c$$

$$Y_1 = \frac{r_c}{\omega^2 L^2} + \frac{1}{j\omega L}$$

$$Y_2 = \frac{1}{R_P} + \frac{1}{j\omega L}$$

Equating Y_1 and Y_2, we get:

$$\frac{r_c}{\omega^2 L^2} + \frac{1}{j\omega L} = \frac{1}{R_P} + \frac{1}{j\omega L}$$

$$\frac{1}{R_P} = \frac{r_c}{\omega^2 L^2}$$

$$= \frac{r_c^2}{r_c\,\omega^2 L^2} = \frac{1}{r_c\,Q_c^2}$$

$$R_p = r_c Q_c^2 = \omega LQ$$

From equation (1), we have:

$$\omega L = Q_c r_c \qquad ...(2)$$

$$R = r_i \,||\, R_p \,||\, r_{b'e} \qquad ...(3)$$

Current gain of the amplifier is given as:

$$A_i = \frac{-g_m R}{1 + j(\omega RC - R/\omega L)} = \frac{-g_m R}{1 + j\omega_o RC(\omega/\omega_o - \omega_o/\omega)} \qquad ...(4)$$

Where,

$$\omega_o^2 = \frac{1}{LC}$$

Q of the tuned circuit at the resonant frequency (ω_o) is given as:

$$Q_1 = \frac{R}{\omega_o L} = \omega_o RC \qquad ...(5)$$

$$A_i = \frac{-g_m R}{1 + jQ_i(\omega/\omega_o - \omega_o/\omega)}$$

At $\omega = \omega_o$, gain is maximum and it is given as:

$$A_{i(max)} = -g_m R \qquad ...(6)$$

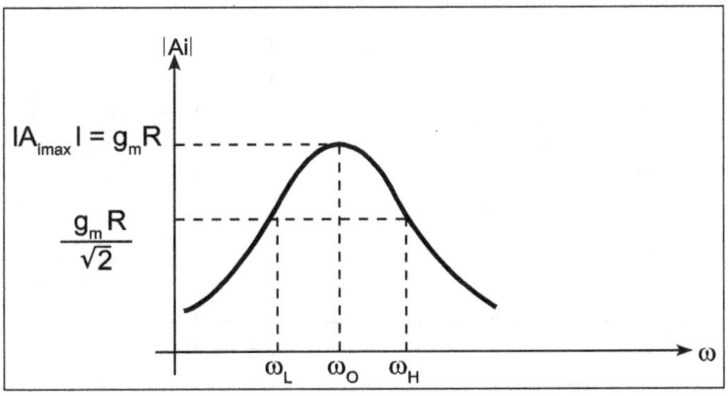

Gain versus frequency for single tuned amplifier.

The above graph shows the variation of the magnitude of the gain as a function of frequency.

At 3 dB frequency, we have:

$$|A_i| = \frac{g_m R}{\sqrt{2}} \quad ...(7)$$

$$1 + jQ_i\left[(\omega/\omega_o) - (\omega_o/\omega)\right] = \sqrt{2}$$

$$1 + Q_i^2\left(\frac{\omega}{\omega_o} - \frac{\omega_o}{\omega}\right)^2 = 2 \quad ...(8)$$

This equation is quadratic in ω_2 and has two positive solutions namely ω_H and ω_L. After solving equation (8), we get 3 dB bandwidth which is given as:

$$BW = \frac{1}{2\pi RC} \quad ...(9)$$

An amplifier that uses a pair of mutually inductively coupled coils where both primary and secondary are tuned is known as double tuned amplifier. Its response will provide substantial rejection of frequencies near the pass band as well as the relatively flat pass band response.

Potential instability is the disadvantage of single tuned amplifiers which we can overcome by using double tuned amplifiers.

Double Tuned Amplifiers

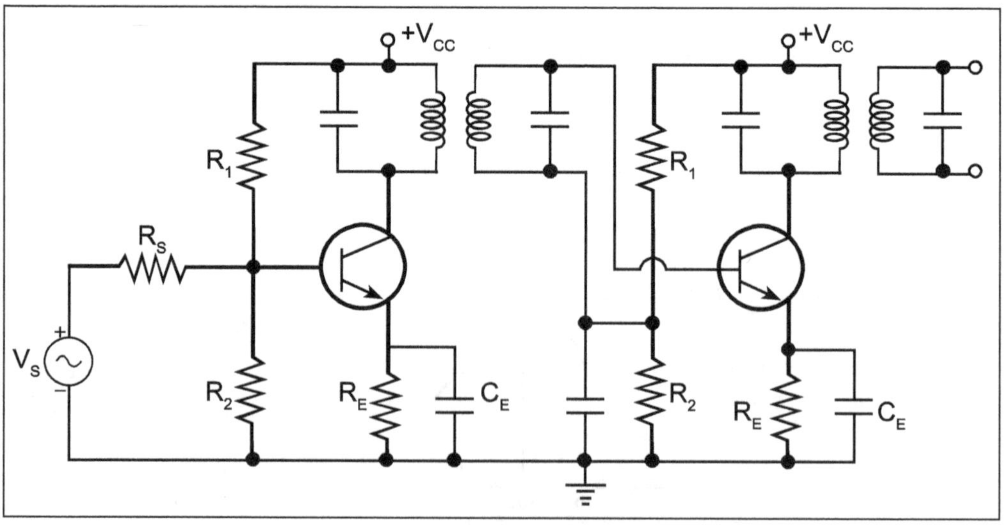

Double tuned amplifier.

It has one or more stages where each stage uses coupled circuits that have different frequencies of resonance. The two resonant circuits are normally inductively coupled. The tuning is done in both secondary and primary circuits.

A double tuned amplifier consists of inductively coupled two tuned circuits. A change in the coupling of the two tuned circuits results in the change in the shape of the frequency response curve.

By proper adjustment of the coupling between the two coils of the two tuned circuits, the required results are obtained.

Operation

The high frequency signal to be amplified is applied to the input terminal of the amplifier. The resonant frequency of tuned circuit connected in the collector circuit is equal to the signal frequency where C_1 varies. Tuned circuit L_1, C_1 offers very high impedance to input signal frequency and large output is developed across it.

The output from tuned circuit L_1,C_1 is transferred to the second tuned circuit L_2, C_2 through mutual induction. Hence, the frequency response in double tuned amplifier depends on the magnetic coupling of L_1 and L_2.

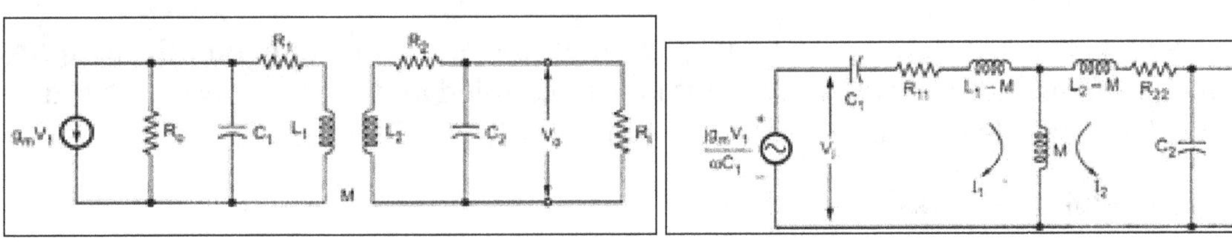

Equivalent circuits of double tuned amplifier.

$$Y_T = \frac{kQ^2}{\omega_r \sqrt{L_1 L_2}\left[4Q\delta - j\left(1 + k^2 Q^2 - 4Q^2 \delta^2\right)\right]}$$

$$|A_v| = g_m \omega_r \sqrt{L_1 L_2}\, Q \frac{kQ}{\sqrt{1 + k^2 Q^2 - 4Q^2 \delta^2 + 16Q^2 \delta^2}}$$

Two gain peaks in frequency f_1 and f_2 is given as:

$$f_1 = f_r\left(1 - \frac{1}{2Q}\sqrt{k^2 Q^2 - 1}\right)$$

$$f_2 = f_r\left(1 + \frac{1}{2Q}\sqrt{k^2 Q^2 - 1}\right)$$

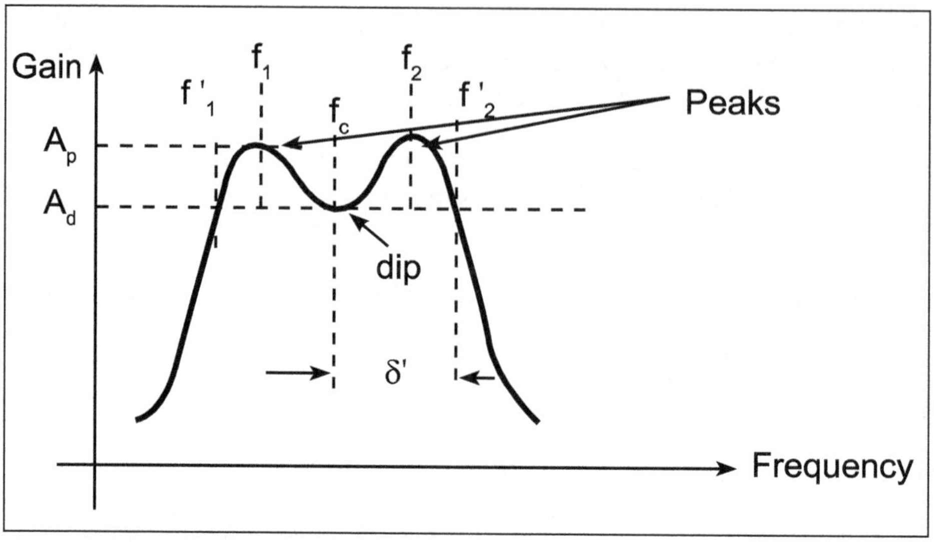

Gain vs frequency curve.

$$k^2 Q^2 = 1, \text{ i.e. } k = \frac{1}{Q}, f_1 = f_2 = f_r$$

The above condition is known as critical coupling.

For the values of $k < 1/Q$, the peak gain is less than the maximum gain and the coupling is poor. For the values $k > 1/Q$, the circuit is over coupled and the response shows double peak. This double peak is useful when more bandwidth is required.

The gain magnitude at peak is given as:

$$|A_p| = \frac{g_m \omega_o \sqrt{L_1 L_2} kQ}{2}$$

Gain at the dip at $\delta = 0$ is given as:

$$|A_d| = |A_p| \frac{2kQ}{1+k^2 Q^2}$$

The ratio of peak and dip gain is denoted as γ and it represents the magnitude of the ripple in the gain curve.

$$\gamma = \left|\frac{A_p}{A_d}\right| = \frac{1+k^2 Q^2}{2kQ}$$

Using quadratic simplification and positive sign, we have:

$$kQ = \gamma + \sqrt{\gamma^2 - 1}$$

Bandwidth is given as:

$$BW = 2\delta = \sqrt{2}(f_2 - f_1)$$

At 3 dB bandwidth, we have:

$$3\,dB\ BW = \frac{3.1 f_r}{Q}$$

Quality factor of a double tuned amplifier is given as:

$$Qeff = fr * (2^{\wedge}(1/n) - 2)^{\wedge}(1/2)/(B.W)$$

Advantages of double tuned amplifier:

- It provides larger 3 dB bandwidth.
- It provides a gain versus frequency curve having steeper sides and flatter tops.
- Double tuned amplifiers use two inductively coupled tuned circuits per stage where both the tuned circuits are tuned to the same frequency.
- Sensitivity can be increased due to increase in the overall gain. Here sensitivity refers to the ability to receive weak signals.
- Selectivity can be increased. This is due to decrease in quality factor and increase in bandwidth. Sensitivity is inversely proportional to quality factor which refers to the ability to discriminate the signals in adjacent band.

4

Transient Analysis

4.1 Natural and Forced Response

Transient Response

A transient response or natural response is the response of a system to a change from equilibrium. The transient response is not necessarily tied to on/off events but to any event that affects the equilibrium of the system.

Transient Time

The time taken for a circuit to change from one steady state to another steady state is called as transient time.

Free and Forced Responses

If the response of a circuit depends on the nature of the circuit, it is called as free or natural response.

If the response of a circuit depends on the source of the circuit, it is called as forced response.

$$i(t) = i_c + i_p = \text{Complementary function} + \text{Particular integral}$$

Where,

i_c - Complementary function dies after sometime is called as natural or transient or source free response [response caused by without source i.e., due to storage elements]

i_p - Particular integral solution is the steady state response or forced response (It is always constant).

Time Constant

The time required for the current to rise to the 0.632 (63.2%) of its final value is known as time constant of the circuit.

Expression of Time Constants of Serious RL and RC Circuit

Circuit	RL circuit	RC circuit
Time constant	$\tau = L/R$ sec	$\tau = RC$ sec

Problems

1. For the circuit shown below, let us determine the current equation where the switch is changed from position 1 to position 2 at t = 0.

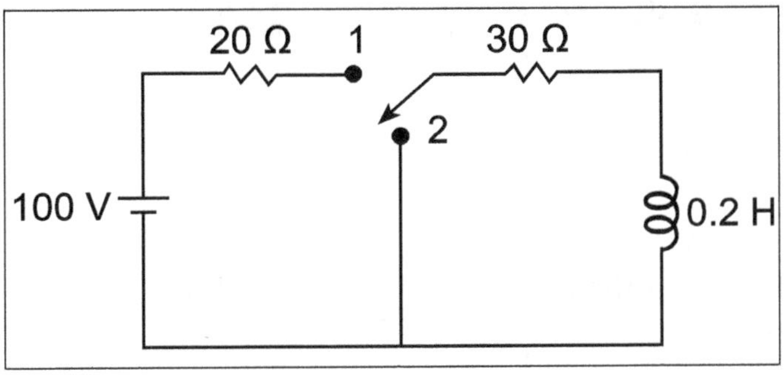

Solution:

Given:

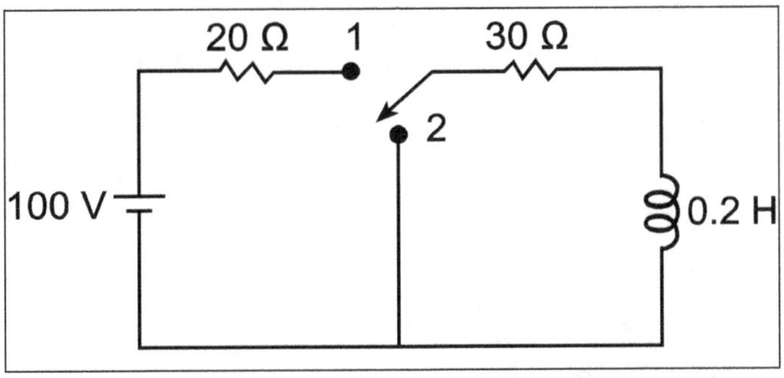

When the switch is at position 2, current equation is written by using Kirchhoff's voltage law as:

$$30i(t)+0.2\frac{di(t)}{dt}=0$$

$$\left(D+\frac{30}{0.2}\right)i=0$$

$$(D+150)i=0$$

$$i=c_1 e^{-150t}$$

At t = 0, the switch is changed to position 2 i.e., i(0) = c1

- At t = 0, the initial current passing through the circuit is same as the current passing through the circuit.
- At t = 0-, the switch is at position 1 and the current passing through the circuit is i = 100/50 = 2 A.

Since the inductor does not allow sudden changes in current, the same current passes through the circuit. Hence, the initial current passing through the circuit when the switch is at position 2 is $i(0^+) = 2A$.

$$C_1 = 2\text{ A}$$

Total current equation is given as:

$$i = 2e^{-150t}$$

2. For the circuit shown in the below figure, let us find the current equation when the switch is opened at t = 0.

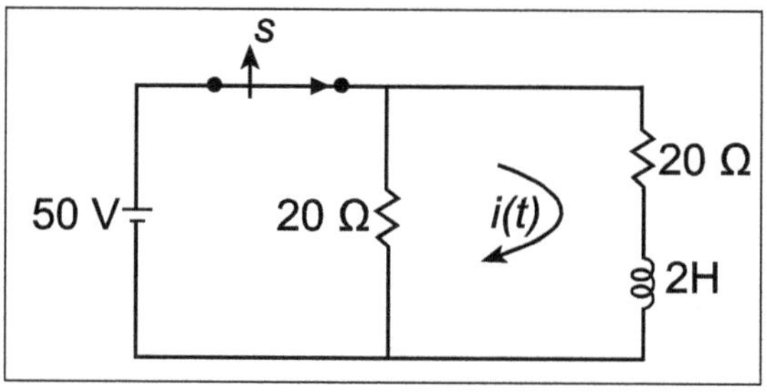

Solution:

Given:

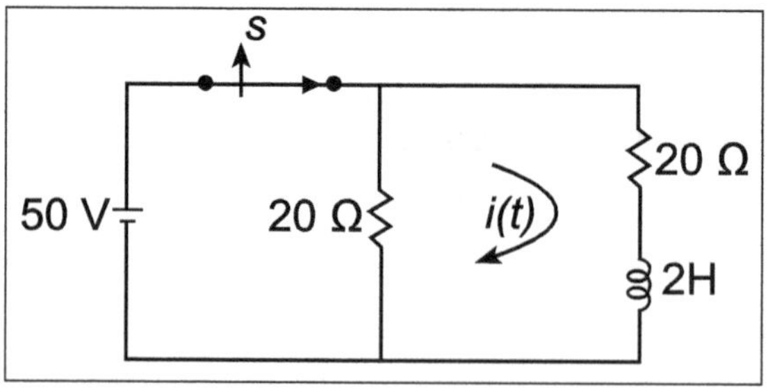

At t = 0, switch S is opened.

By using Kirchhoff's voltage law, the current equation is written as:

$$20i+20i+2\frac{di}{dt}=0$$

$$40i+2\frac{di}{dt}=0$$

$$D + 20i = 0$$

The solution for the above equation is given as:

$$i = c_1 e^{-20t}$$

When the switch is closed for a time, the inductor acts as short circuit for dc voltages where the current passing through the inductor is 2.5 A.

Before the switch is opened, the current passing through the inductor is 2.5 A. Since the current in the inductor cannot change instantaneously, $i(0^+)$ is equal to 2.5 A.

$$\text{At } t = 0,\ c_1 = i\left(0^+\right) = 2.5$$

Final solution is given as:

$$i(t) = 2.5\,e^{-20t}$$

3. For the circuit shown below, let us determine the current equation when the switch is opened at t = 0.

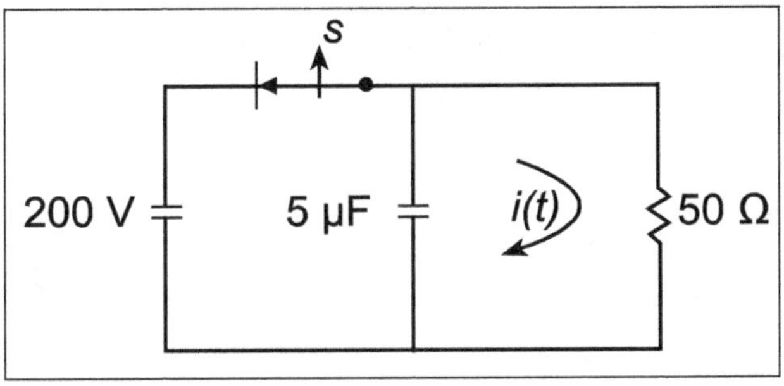

Solution:

Given:

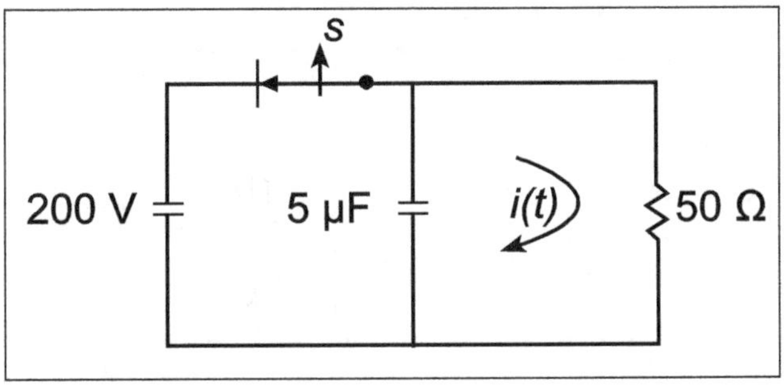

By using Kirchhoff's voltage law, the current equation is given as:

$$\frac{1}{5\times10^{-6}}\int idt+50i=0$$

Differentiating the equation once, we have:

$$50\frac{di}{dt}+\frac{1}{5\times10^{-6}}i=0$$

$$\therefore\left(D+\frac{1}{250\times10^{-6}}\right)i=0$$

$$\therefore i=c_1\exp\left(\frac{-1}{250\times10^{-6}}t\right)$$

At t = 0-, just before the switch S is opened, the voltage across the capacitor is 200V. Since the voltage across the capacitor cannot change instantly, it remains equal to 200V at t = 0⁺. At that instant, the current through the resistor is given as:

$$i\left(0^+\right)=\frac{200}{50}=4A$$

In the above equation, the current equation is $i(0^+)$ at t = 0.

$$c_1=4A$$

The current equation is given as:

$$i=4\exp\left(\frac{-1}{250\times10^{-6}}t\right)A$$

4. For the circuit shown below, let us find the current equation when the switch S is opened at t = 0.

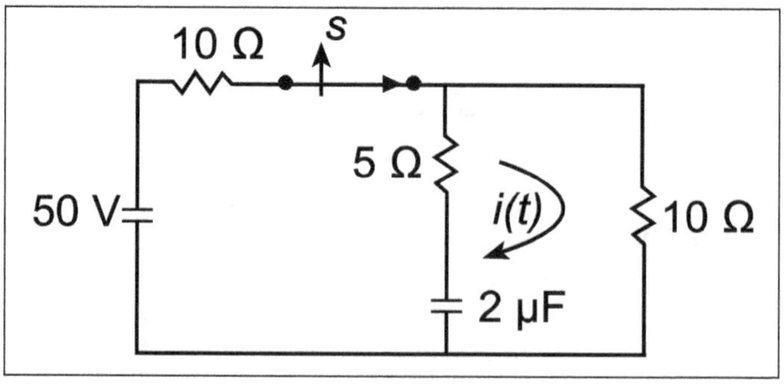

Solution:

Given:

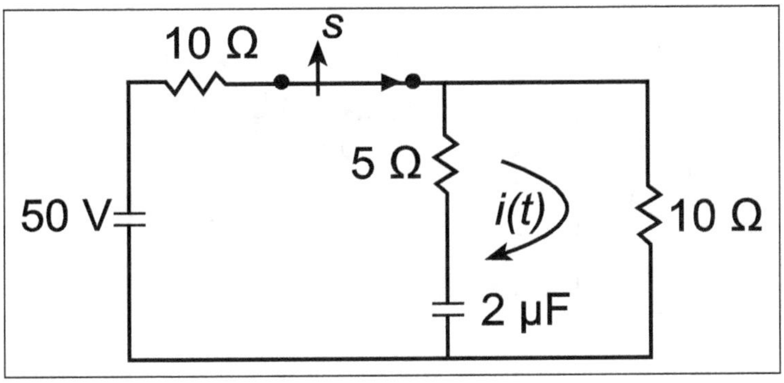

By using Kirchhoff's voltage law, the current equation is given as:

$$\frac{1}{2\times10^{-6}}\int i\,dt+5i+10i=0$$

Differentiating the above equation, we get:

$$15\frac{di}{dt}+\frac{i}{2\times10^{-6}}=0$$

$$\left(D+\frac{1}{30\times10^{-6}}\right)i=0$$

$$i=c_1\exp\left(\frac{-1}{30\times10^{-6}}\right)t$$

At t = 0^-, just before switch S is opened, the current through 10Ω resistor is 2.5A. The same current passes through 10 Ω at t = 0^+.

$$\therefore\ i\left(0^+\right)=2.5A$$

$$\text{At } t=0,\ \ i\left(0^+\right)=2.5A$$

$$\therefore\ c_1=2.5$$

Complete solution is given as:

$$i\ =\ 2.5\exp\left(\frac{-1}{30\times10^{-6}}t\right)$$

5. For the circuit shown below, let us determine the complete expression for the current when the switch is closed at t = 0.

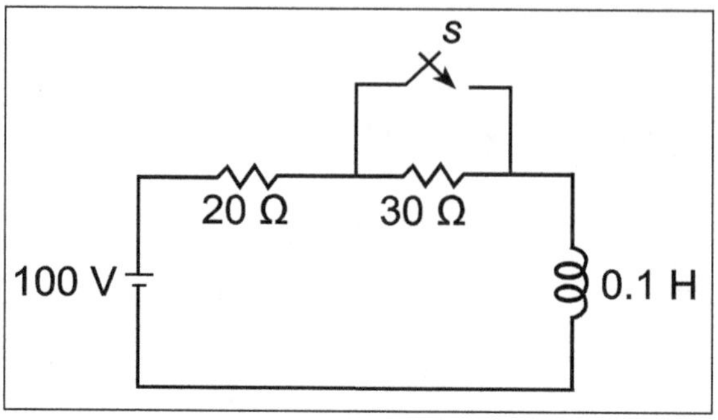

Solution:

Given:

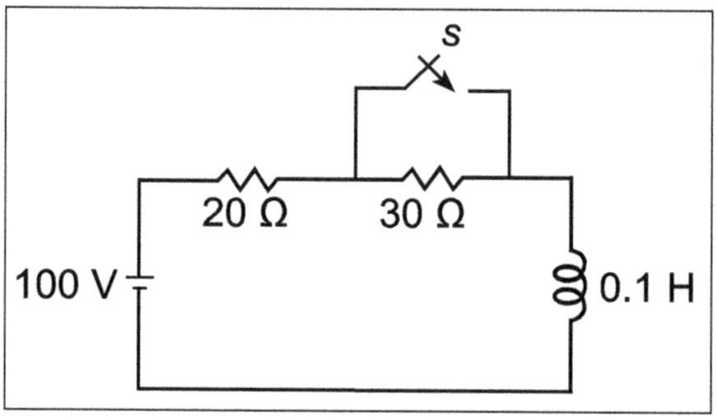

By using Kirchhoff's voltage law, the differential equation when the switch is closed at t = 0 is given as:

$$20i+0.1\frac{di}{dt}=100$$

$$(D + 200)\,i = 1000$$

$$i=c_1 e^{-200t}+e^{-200t}\int 1000e^{200t}\,dt$$

$$\therefore\ i=c_1 e^{-200t}+5$$

At t = 0-, the current passing through the circuit is $i(0^-)=\frac{100}{50}=2A$.

Since, the inductor does not allow sudden changes in currents at t = 0^+, the same current passes through circuit.

$$\therefore\ i(0^+)=2A$$

At $t = 0,\ i(0^+) = 2$

$\therefore\ c_1 = -3$

The complete solution is given as:

$$i = -3\,e^{-200t} + 5\ A$$

6. The circuit shown below consists of series RL elements where R = 150Ω and L = 0.5 H. The switch is closed when φ = 30°. Let us determine the resultant current when voltage V = 50 cos (100t + φ) is applied to the circuit at φ = 30°.

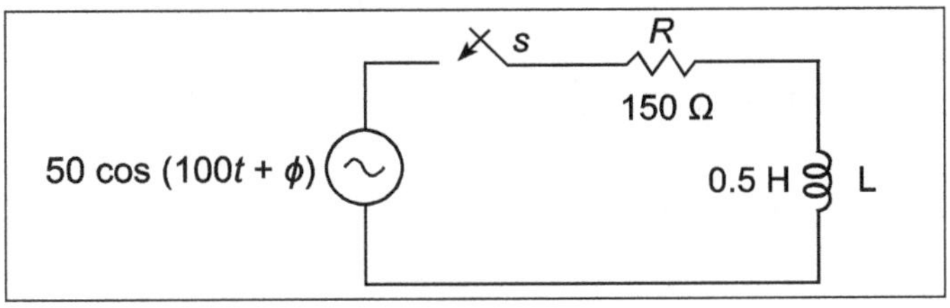

Solution:

Given:

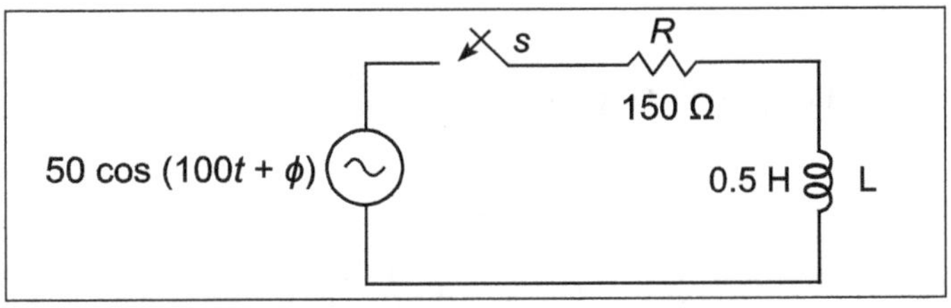

By using Kirchhoff's voltage law, the differential equation when the switch is closed at φ = 30° is given as:

$$150i + 0.5\frac{di}{dt} = 50\cos(100t + \phi)$$

$$0.5\ Di\ +\ 150i = 50\ \cos\ (100t\ +\ 30°)$$

$$(D + 300)i = 100\ \cos\ (100t + 30°)$$

The complementary current is given as:

$$i_c = ce^{-300t}$$

$$i_p = A\cos\ (100t + 30°)$$

$$i_p' = -100\,A \sin(100t + 30°) + 100\,B \cos(100t + 30°)$$

Substituting i_p and i_p' in the differential equation and equating the coefficients, we get as:

$$-100\,A \sin(100\,t + 30°) + 100\,B \cos(100\,t + 30°) + 300\,A \cos(100\,t + 30°)$$
$$+300\,B \sin(100\,t + 30°) = 100 \cos(100\,t + 30°) \;\; -100\,A + 300$$

$$A = 0.3 \text{ and } B = 0.1$$

The particular current is given as:

$$i_p = 0.3 \cos(100\,t + 30°) + 0.1 \sin(100t + 30°)$$

$$i_p = 0.316 \cos(100\,t + 11.57°)\ A$$

Complete equation for the current is given as:

$$i = i_p + i_c$$

$$i = ce^{-300t} + 0.316 \cos(100\,t + 11.57°)$$

At t = 0, current $i_o = 0$

$$c = -0.316 \cos(11.57°) = -0.309$$

Complete solution for the current is given as:

$$i = -0.309\,e^{-300t} + 0.316 \cos(100\,t + 11.57°)\ A$$

7. For the series parallel circuit shown in the below figure, let us determine the following.

(a) Total impedance between the terminals a, b and state if it is inductive or capacitive.

(b) Voltage across the parallel branch.

(c) Phase angle.

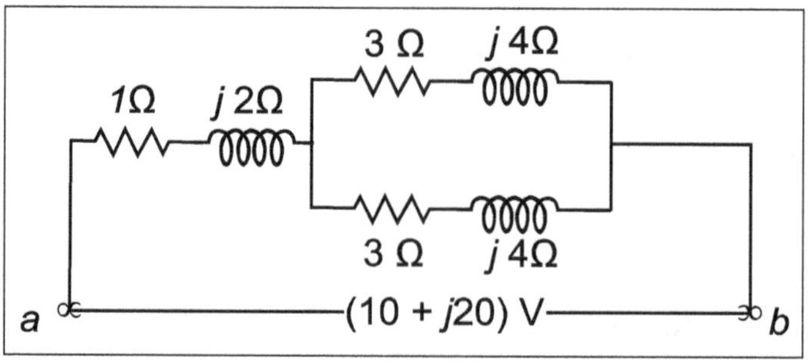

Solution:

Given:

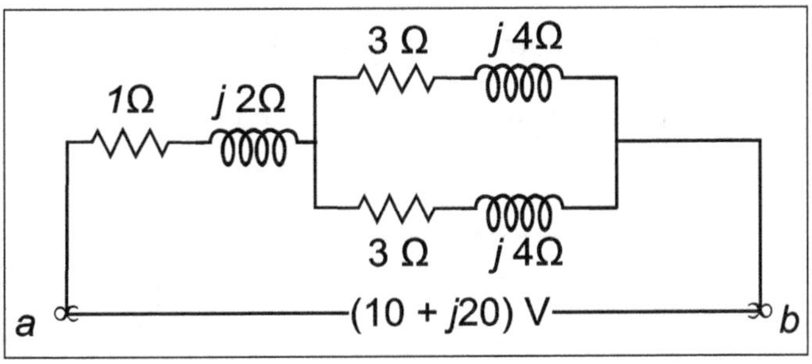

Parallel branch is combined into a single branch.

$$Z_P = (3+j4)\|(3+j4) = (1.5+j2)\Omega$$

Total impedance is given as,

$$Z_T = 1+j2+1.5+j2 = (2.5+j4)\Omega$$

Total impedance in the circuit is inductive.

Total current in the circuit is given as:

$$I_T = \frac{V_S}{Z_T} = \frac{10+j20}{2.5+j4}$$

$$= \frac{22.36\angle 63.43^\circ}{4.72\angle 57.99^\circ}$$

$$I_T = 4.72\angle 5.44^\circ \text{ A}$$

Current lags the voltage by 57.99°.

Phase angle is given as:

$$\theta = 57.99^\circ$$

Voltage across the parallel branch is given as:

$$V_P = (1.5+j2)4.74\ \angle 5.44^\circ$$

$$= 2.5 \times 4.74\angle(5.44^\circ + 53.13^\circ)$$

$$= 11.85\angle 58.57^\circ \text{V}$$

4.2 Transient Response of RC, RL and RLC Circuits to Excitation

RC, RL, RLC circuits using Laplace transform for DC input:

DC response of RL circuit with initial condition of i(0) = 0

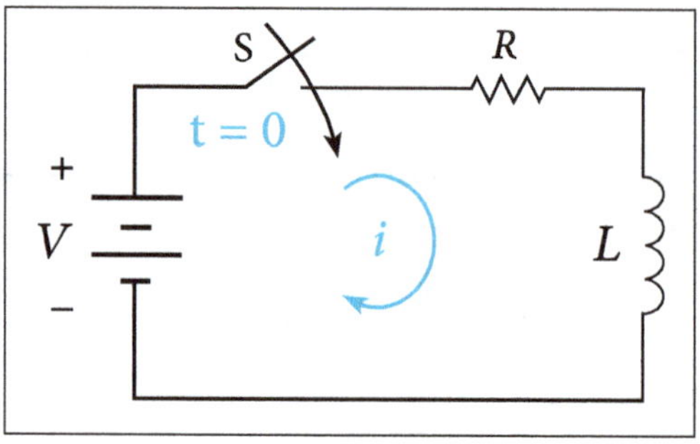

By applying KVL, we have:

$$V = Ri(t) + L\frac{di}{dt}$$

Taking Laplace transforms on both sides, we have:

$$\frac{V}{s} = RI(s) + L\left[s\,I(s) - i(0)\right]$$

$$\frac{V}{s} = [R + sL]I(s)$$

$$I(s) = \frac{V}{s[sL + R]} = \frac{V}{L} * \frac{1}{s(s + R/L)}$$

By using Laplace transform, we have:

$$\frac{1}{s(s + R/L)} = \frac{A}{s} + \frac{B}{s + R/L}$$

$$1 = A(s + R/L) + Bs$$

Substituting s = 0, we have:

$$1 = A(R/L) \rightarrow A = \frac{L}{R}$$

$$s=-\frac{R}{L},1=B(-R/L)\Rightarrow B=-\frac{L}{R}$$

$$I(s)=\frac{V}{L}*\left[\frac{L/R}{s}-\frac{L/R}{s+\frac{R}{L}}\right]$$

$$I(s)=\frac{V}{L}\left[\frac{1}{s}-\frac{1}{s+R/L}\right]$$

Taking Laplace inverse transform, we have:

$$i(t)=\frac{V}{R}\left(1-e^{-Rt/L}\right)$$

Time Response of RL Circuit with DC Input

The current equation for RL circuit is given as:

$$i(t)=\frac{V}{R}\left(1-e^{-Rt/L}\right)$$

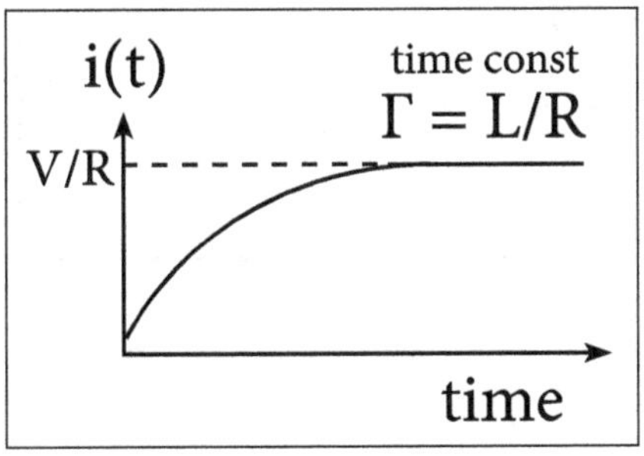

(a) Voltage across resistor:

$$V_R=R*i=R\frac{V}{R}\left(1-e^{\frac{-Rt}{L}}\right)$$

$$V_R=V\left[1-e^{\frac{-Rt}{L}}\right]$$

(b) Voltage across inductor:

$$V_L=L\frac{di}{dt}$$

$$V_L = L\frac{d}{dt}\left(\frac{V}{R}\left[1-e^{\frac{-Rt}{L}}\right]\right)$$

$$V_L = L\frac{V}{R}\left[-e^{\frac{-Rt}{L}}\right]*\left(\frac{-R}{L}\right) = Ve^{\frac{-Rt}{L}}$$

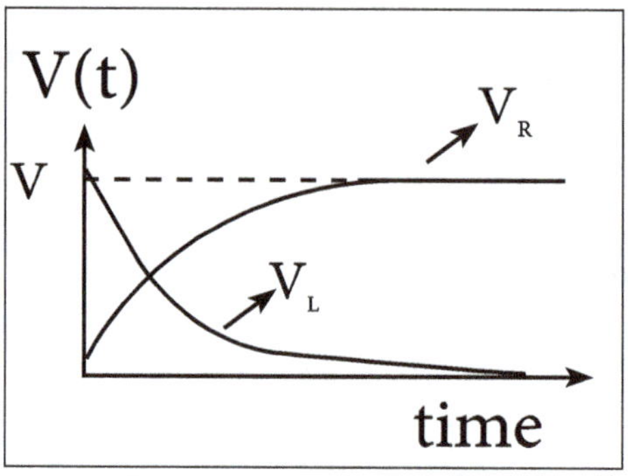

Time response of V_R and V_L.

DC response of RC circuit with initial voltage across capacitor v (0) = 0

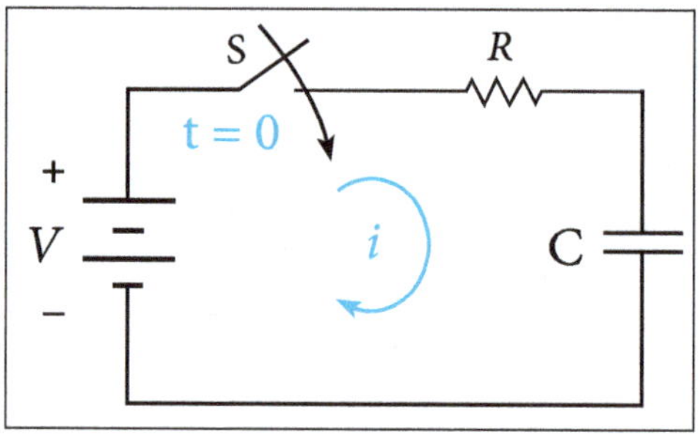

By applying KVL, we have:

$$V = Ri(t) + \frac{1}{C}\int i\,dt$$

Taking Laplace transform on both sides, we have:

$$\frac{V}{s} = RI(s) + \frac{1}{Cs}I(s)$$

$$\frac{V}{s} = \left[R + \frac{1}{Cs}\right]I(s)$$

$$\frac{V}{s} = \left[\frac{RCs+1}{Cs}\right]I(s)$$

$$I(s)=\frac{VC}{[RCs+1]}$$

Taking RC commonly out from denominator, we have:

$$I(s)=\frac{VC}{RC}*\frac{1}{(s+1/RC)}$$

$$I(s)=\frac{V}{R}*\frac{1}{(s+1/RC)}$$

By using Laplace inverse transform, we have:

$$i(t)=\frac{V}{R}*e^{-t/RC}$$

Laplace transform is given as:

$$\frac{1}{S+a}\leftrightarrow e^{-at}$$

$$i(t)=\frac{V}{R}e^{-t/RC}$$

Time Response of RC Circuit with DC Input

Current equation for RC circuit is given as:

$$i(t)=\frac{V}{R}e^{-t/RC}$$

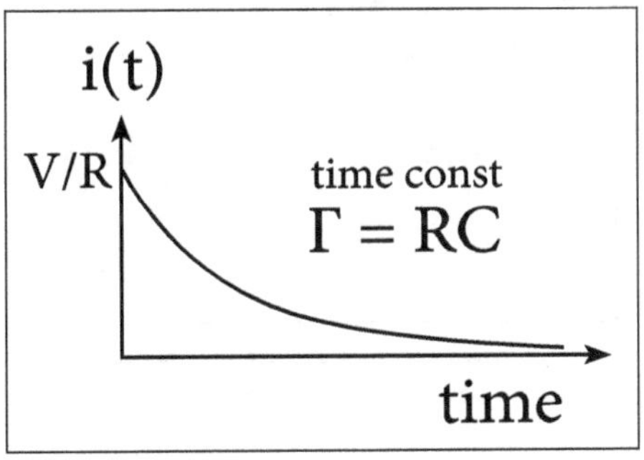

(a) Voltage across resistor:

$$V_R=R*i(t)=R\frac{V}{R}e^{\frac{-t}{RC}}$$

$$V_R = Ve^{\frac{-t}{RC}}$$

(b) Voltage across capacitor:

$$V_C = \frac{1}{C}\int_o^t i\,dt$$

$$V_C = \frac{1}{C}\int_o^t \frac{V}{R} e^{-\frac{t}{RC}} dt$$

$$V_C = \frac{V}{RC}\left[-RCe^{-\frac{t}{RC}}\right]_o^t = V\left[1-e^{-\frac{t}{RC}}\right]$$

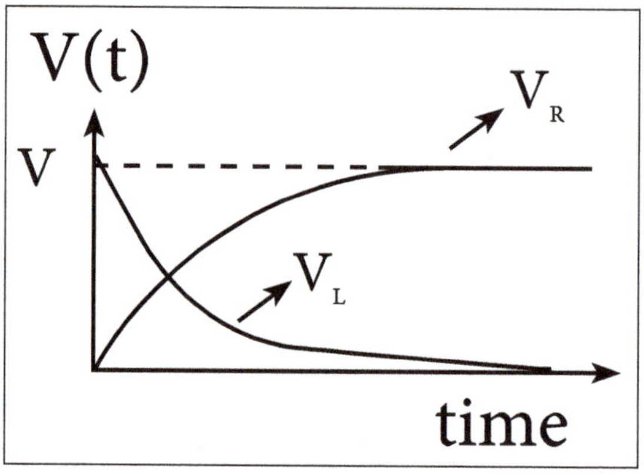

Time response of V_R and V_L.

DC response of RLC circuit where initial conditions are equal to zero

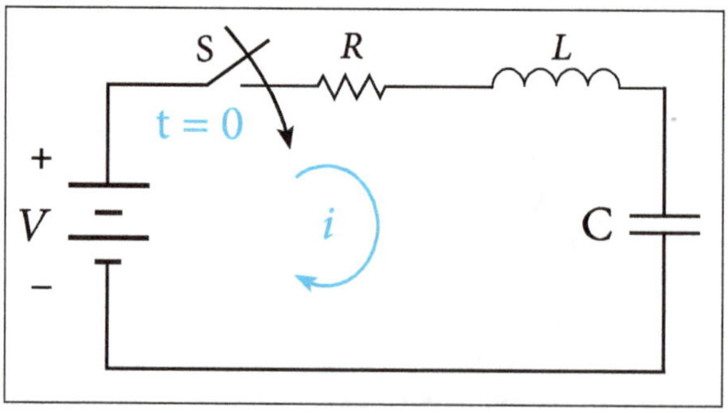

Series RLC circuit.

By applying KVL, we have:

$$V = Ri + L\frac{di}{dt} + \frac{1}{C}\int i\,dt$$

Differentiating on both sides, we have:

$$0=R\frac{di}{dt}+L\frac{d^2 i}{dt^2}+\frac{1}{C}i$$

Re-arranging the above equation, we have:

$$\frac{d^2 i}{dt^2}+\frac{R}{L}\frac{di}{dt}+\frac{1}{LC}i=0$$

Above equation can be written as:

$$D^2+\frac{R}{L}D+\frac{1}{LC}=0$$

By solving the above equation, we have:

$$D_1,D_2=-\frac{R}{2L}\pm\sqrt{\left(\frac{R}{2L}\right)^2-\frac{1}{LC}}$$

$$D_1,D_2 \Rightarrow K_1 \pm K_2$$

$$K_1=-\frac{R}{2L},K_2=\sqrt{\left(\frac{R}{2L}\right)^2-\frac{1}{LC}}$$

Based on K_2 value, the solution becomes:

Case (i):

If K_2 is positive, then the roots become real which gives over damped response.

$$i(t)=c_1 e^{(K_1+K_2)t}+c_2 e^{(K_1-K_2)t}$$

Case (ii):

If K_2 is zero, then the roots become real which gives critically damped response.

$$i(t)=e^{K_1 t}\left[c_1 t+c_2\right]$$

Case (iii):

If K_2 is negative, then the roots become complex conjugate which gives under damped response.

$$i(t)=e^{K_1 t}\left[c_1 \cos K_2 t+c_2 \sin K_2 t\right]$$

Time Response of RLC Circuit with DC Input

Case (i):

If K_2 is positive, then the roots become real which gives over damped response.

$$i(t)=c_1 e^{(K_1+K_2)^t}+c_2 e^{(K_1-K_2)^t}$$

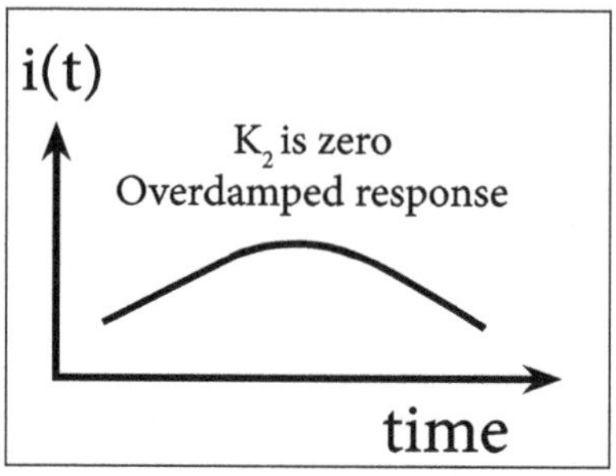

Case (ii):

If K_2 is zero, then the roots become real which gives critically damped response.

$$i(t)=e^{K_1 t}\left[c_1 t+c_2\right]$$

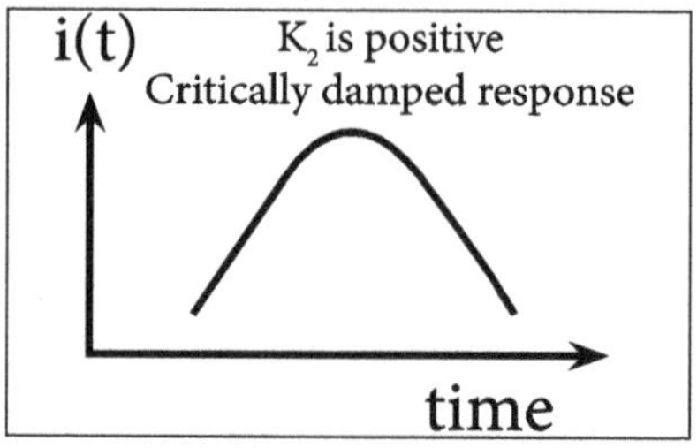

Case (iii):

If K_2 is negative, then the roots become complex conjugate which gives under damped response.

$$i(t)=e^{K_1 t}\left[c_1 \cos K_2 t+c_2 \sin K_2 t\right]$$

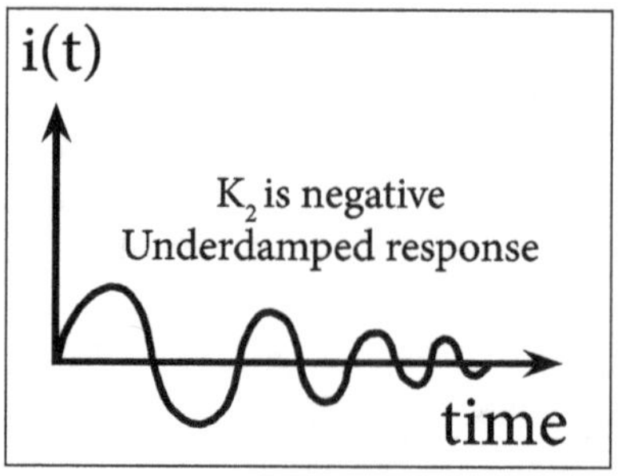

RC, RL, RLC Circuits using Laplace Transform for AC or Sinusoidal Input

(a) AC or Sinusoidal Response of RL Circuit

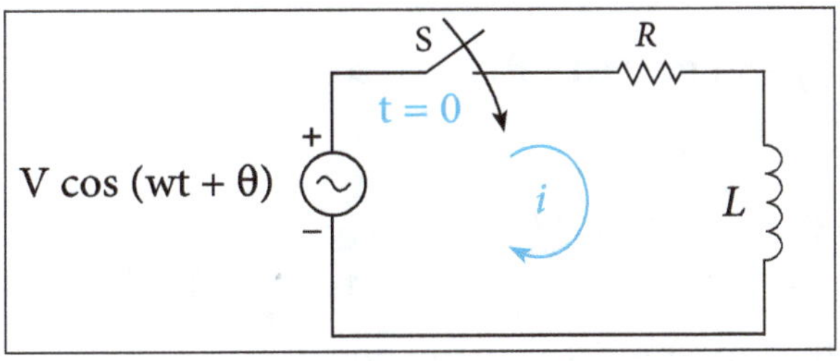

Step (1):

By applying KVL, we have:

$$Ri(t)+L\frac{di(t)}{dt}=V\cos(wt+\theta)$$

Divide both sides by L, then:

$$\frac{di(t)}{dt}+\frac{R}{L}i(t)=\frac{V}{L}\cos(wt+\theta) \quad ...(1)$$

Complete solution is given as:

$$i(t)=i_c+i_p \quad ...(2)$$

Step (2):

To find i_c, RHS of equation (1) is equated to 0.

$$\frac{di(t)}{dt}+\frac{R}{L}i(t)=0$$

$$\left(D+\frac{R}{L}\right)i(t)=0$$

$$i_c=Ce^{-RT/L}$$

Step (3):

To find particular integral i_p, we have:

$$i_p=A\cos(wt+\theta)+B\sin(wt+\theta)$$

$$\frac{di_p}{dt}(\text{or})i_p = -Aw\sin(wt+\theta) + Bw\cos(wt+\theta)$$

Substituting i_p and i_p' in the above equation (1), then:

$$-Aw\sin(wt+\theta) + Bw\cos(wt+\theta) + \frac{R}{L}A\cos(wt+\theta)$$

$$+\frac{R}{L}B\sin(wt+\theta) = \frac{V}{L}\cos(wt+\theta)$$

Equating cos and sin term on both sides, we have:

$$\left.\begin{aligned} A\frac{R}{L} + Bw = \frac{V}{L} \\ -Aw + B\frac{R}{L} = 0 \end{aligned}\right\} \text{by solving this,}$$

$$A = \frac{VR}{R^2 + (wL)^2}$$

$$B = \frac{V(wL)}{R^2 + (wL)^2}$$

$$A = M\cos\Phi$$

$$B = M\sin\Phi$$

$$i_p = M\cos\Phi\cos(wt+\theta) + M\sin\Phi\sin(wt+\theta)$$

Re-writing the above equation:

$$i_p = M\cos(wt+\theta)\cos\Phi + M\sin(wt+\theta)\sin\Phi$$

$$\cos A\cos B + \sin A\sin B = \cos(A-B)$$

$$i_p = M\cos(wt+\theta-\Phi) \quad ...(3)$$

$$A = M\cos\Phi$$

$$B = M\sin\Phi$$

$$M^2\cos^2\Phi + M^2\sin^2\Phi = A^2 + B^2$$

$$M^2\left[\cos^2\Phi+\sin^2\Phi\right]=\frac{V^2R^2}{\left(R^2+(wL)^2\right)^2}+\frac{V^2(wL)^2}{\left(R^2+(wL)^2\right)^2}$$

$$M^2=\frac{V^2R^2+(wL)^2}{\left(R^2+(wL)^2\right)^2}$$

$$\sqrt{R \quad (wL)}$$

$$\frac{M\sin\Phi}{M\cos\Phi}=\frac{B}{A}\Rightarrow\tan\Phi=\left(\frac{wL}{R}\right)$$

$$\Phi=\tan^{-1}\left(\frac{wL}{R}\right)$$

Step (4):

By substituting M and ϕ in the above equation (3), we have:

$$i_p=\frac{V}{\sqrt{R^2+(wL)^2}}\cos\left[wt+\theta-\tan^{-1}\left(\frac{wL}{R}\right)\right]$$

Step (5):

Complete solution becomes:

$$i(t)=i_c+i_p$$

$$i(t)=ce^{-RT/L}+\frac{V}{\sqrt{R^2+(wL)^2}}\cos\left[wt+\theta-\tan^{-1}\left(\frac{wL}{R}\right)\right]$$

(b) AC or Sinusoidal Response of RC Circuit

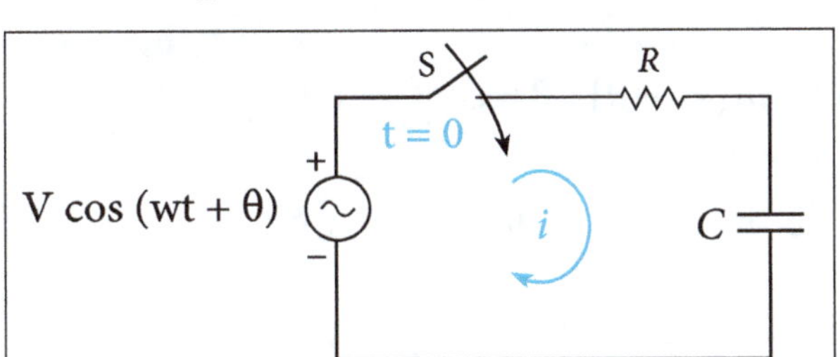

Step (1):

By applying KVL, we have:

$$Ri(t)+\frac{1}{C}\int i(t)dt = V\cos(wt+\theta)$$

Differentiating on both sides, we have:

$$R\frac{di(t)}{dt}+\frac{1}{C}i(t)=-Vw\sin(wt+\theta)$$

Dividing both sides by R, then:

$$\frac{di(t)}{dt}+\frac{1}{RC}i(t)=-\frac{Vw}{R}\sin(wt+\theta) \quad ...(1)$$

Complete solution is given as:

$$i(t)=i_c+i_p \quad ...(2)$$

Step (2):

To find i_c, RHS of equation (1) is equated to 0.

$$\frac{di(t)}{dt}+\frac{1}{RC}i(t)=0$$

$$\left(D+\frac{1}{RC}\right)i(t)=0$$

$$i_c=Ce^{-t/RC}$$

Step (3):

To find particular integral i_p, we have:

$$i_p=A\cos(wt+\theta)+B\sin(wt+\theta)$$

$$\frac{di_p}{dt}(\text{or})\,i_p=Aw\sin(wt+\theta)+Bw\cos(wt+\theta)$$

Substituting i_p and i_p' in the above equation (1), we have:

$$-Aw\sin(wt+\theta)+Bw\cos(wt+\theta)+\frac{}{RC}A\cos(wt+\theta)$$

$$+\frac{1}{RC}B\sin(wt+\theta)=-\frac{Vw}{R}\sin(wt+\theta)$$

Equating cos and sin term on both sides, we have:

$$\left.\begin{aligned} A\frac{1}{RC}+Bw=0 \\ -Aw+B\frac{1}{RC}=-\frac{Vw}{R} \end{aligned}\right\}\text{by solving this,}$$

$$A=\frac{VR}{R^2+(1/wC)^2}$$

$$B=\frac{-V(1/wC)}{R^2+(1/wC)^2}$$

$$A=M\cos\Phi$$

$$B=M\sin\Phi$$

$$i_p=M\cos\Phi\cos(wt+\theta)+M\sin\Phi\sin(wt+\theta)$$

Re-writing the above equation, we have:

$$i_p=M\cos(wt+\theta)\cos\Phi+M\sin(wt+\theta)\sin\Phi$$

$$\cos A\cos B+\sin A\sin B=\cos(A-B)$$

$$i_p=M\cos(wt+\theta-\Phi)\quad ...(3)$$

$$A=M\cos\Phi$$

$$B=M\sin\Phi$$

$$M^2\cos^2\Phi+M^2\sin^2\Phi=A^2+B^2$$

$$M^2\left[\cos^2\Phi+\sin^2\Phi\right]=\frac{V^2R^2}{\left(R^2+(1/wC)^2\right)^2}+\frac{V^2(1/w/C)^2}{\left(R^2+(wL)^2\right)^2}$$

$$M^2=\frac{V^2R^2+(1/wC)^2}{\left(R^2+(1/wC)^2\right)^2}$$

$$M = \frac{V}{\sqrt{R^2 (1/wC)^2}}$$

$$\frac{M\sin\Phi}{M\cos\Phi} = \frac{B}{A} \Rightarrow \tan\Phi = -\left(\frac{1/wC}{R}\right)$$

$$\Phi = \tan^{-1}\left(\frac{1}{wCR}\right)$$

Step (4):

By substituting M and ϕ in the above equation (3), we have:

$$i_p = \frac{V}{\sqrt{R^2 + (1/wC)^2}} \cos\left[wt + \theta + \tan^{-1}\left(\frac{1}{wCR}\right)\right]$$

Step (5):

Complete solution is given as:

$$i(t) = i_c + i_p$$

$$i(t) = ce^{-t/RC} + \frac{V}{\sqrt{R^2 + (1/wC)^2}} \cos\left[wt + \theta + \tan^{-1}\left(\frac{1}{wCR}\right)\right]$$

(c) AC or Sinusoidal Response of RLC Circuit

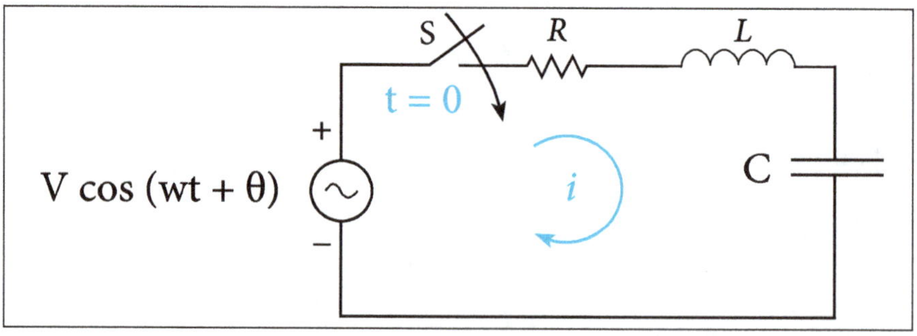

Step (1):

By applying KVL, we have:

$$Ri(t) + L\frac{di(t)}{dt} + \frac{1}{C}\int i(t)dt = V\cos(wt + \theta)$$

Differentiating on both sides, we have:

$$R\frac{di(t)}{dt}+L\frac{d^2i(t)}{dt^2}+\frac{1}{C}i(t)=-Vw\sin(wt+\theta)$$

Dividing both sides by L, we have:

$$\frac{d^2i(t)}{dt^2}+\frac{R}{L}\frac{di(t)}{dt}+\frac{1}{LC}i(t)=-\frac{Vw}{L}\sin(wt+\theta) \quad ...(1)$$

Complete solution is given as:

$$i(t)=i_c+i_p \quad ...(2)$$

Where:

i_C = Complementary integral

i_P = Particular integral

Step (2):

To find i_c, RHS of equation (1) is equated to 0.

$$\frac{d^2i(t)}{dt^2}+\frac{R}{L}\frac{di(t)}{dt}+\frac{1}{LC}i(t)=0$$

$$\left(D^2+\frac{R}{L}D+\frac{1}{LC}\right)i(t)=0$$

Based on K_2 value, we have the following three cases.

Case (i):

If K_2 is positive, then the roots become real and different which gives over damped response.

$$i(t)=c_1e^{(K_1+K_2)t}+c_2e^{(K_1-K_2)t}$$

Case (ii):

If K_2 is zero, then the roots become real and same which gives critically damped response.

$$i(t)=e^{K_1t}\left[c_1t+c_2\right]$$

Case (iii):

If K_2 is negative, then the roots become complex conjugate which gives under damped response.

$$i(t)=e^{K_1 t}\left[c_1 \cos K_2 t+c_2 \sin K_2 t\right]$$

Step (3):

Particular integral

$$i_p = A\cos(wt+\theta)+B\sin(wt+\theta)$$

$$\frac{di_p}{dt}=(or)i_p=-A\,w\sin(wt+\theta)+B\,w\cos(wt+\theta)$$

$$\frac{d^2 t}{dt^2}=-Aw^2\cos(wt+\theta)-B\,w^2\sin(wt+\theta)$$

Substituting i_p, i_p' and i_p" in the above equation (1), we have:

$$-Aw^2\cos(wt+\theta)-Bw^2\sin(wt+\theta)+\frac{R}{L}\left[-Aw\sin(wt+\theta)+B\,w\cos(wt+\theta)\right]$$

$$+\frac{1}{RC}\left[A\cos(wt+\theta)+B\sin(wt+\theta)\right]=-\frac{Vw}{L}\sin(wt+\theta)$$

By equating cos and sin term on both sides, we have:

$$\left.\begin{aligned} -Aw^2+Bw\frac{R}{L}+A\frac{1}{RC}&=0 \\ -Bw^2-A\frac{R}{L}+B\frac{1}{RC}&=\frac{Vw}{L} \end{aligned}\right\}\text{by solving this,}$$

$$A=\frac{VR}{R^2+\left(\frac{1}{wC}-wL\right)^2};B=\frac{-V\left(\frac{1}{wC}-wL\right)}{R^2+\left(\frac{1}{wC}-wL\right)^2}$$

$$A = M\cos\Phi$$

$$B = M\sin\Phi$$

$$i_p = M\cos\Phi\cos(wt+\theta)+M\sin\Phi\sin(wt+\theta)$$

Re-writing the above equation, we have:

$$i_p = M\cos(wt+\theta)\cos\Phi + M\sin(wt+\theta)\ \sin q\Phi$$

$$\cos A\cos B\ +\ \sin A\sin B\ =\ \cos(A\ -\ B)$$

$$i_p = M\cos(wt+\theta-\theta)\ ...(3)$$

$$A = M\ \cos\Phi$$

$$B = M\ \sin\Phi$$

$$M = \frac{V}{\sqrt{R^2 + \left(\frac{1}{wC - wL}\right)^2}};\ \Phi = \tan^{-1}\left(\frac{1}{wCR} - \frac{wL}{R}\right)$$

Step (4):

By substituting M and ϕ in equation (3), we have:

$$i_p = \frac{V}{\sqrt{R^2 + \left(\frac{1}{wC} - wL\right)^2}}\cos\left[\ wt+\theta+\tan^{-1}\left(\frac{1}{wCR} - \frac{wL}{R}\right)\right]$$

Step (5):

Complete solution is given as:

$$i(t) = i_c + i_p$$

Second Order Differential Equation

A second order ordinary differential equation has the general form as follows:

$$y'' = f(t, y, y')$$

Where,

f = Function

This equation is said to be linear if 'f' is linear in y and y'. Otherwise, the equation is said to be nonlinear.

$$y'' = g(t) - p(t)y' - q(t)y$$

A second order linear equation is given as:

$$P(t)y''+Q(t)y'+R(t)y=G(t)$$

If G (t) = 0 for all t, then the equation is called as homogeneous. Otherwise, the equation is known as non-homogeneous.

Generation of a Differential Equation in an Output Variable

Output equation is given as:

$$y = Cx + Du$$

$$y(t)=\begin{bmatrix} c_1 & c_2 \end{bmatrix}\begin{bmatrix} x_1 \\ x_2 \end{bmatrix}+[d]u(t)$$

$$=c_1 x_1(t)+c_2 x_2(t)+du(t) \qquad ...(1)$$

In the Laplace domain, we have:

$$Y(s)=\left(C(sI-A)^{-1}B+D\right)U(s)$$

$$=\frac{1}{\det[sI-A]}\left(C\operatorname{adj}(sI-A)+\det[sI-A]D\right)$$

The determinants may be expanded and the resulting equation is written as a differential equation as follows.

$$\frac{d^2y}{dt^2}-(a_{11}+a_{22})\frac{dy}{dt}+(a_{11}a_{22}-a_{12}a_{21})y=q_2\frac{d^2u}{dt^2}+q_1\frac{du}{dt}+q_0u \qquad ...(2)$$

In terms of the standard system parameters, we have:

$$\frac{d^2y}{dt^2}+2\zeta\omega_n\frac{dy}{dt}+\omega_n^2\, y=q_2\frac{d^2u}{dt^2}+q_1\frac{du}{dt}+q_0u \qquad ...(3)$$

Coefficients q_0, q_1 and q_2 are given as,

$$q_0=c_1(-a_{22}b_1+a_{12}b_2)+c_2(-a_{11}b_2+a_{21}b_1)+d(a_{11}a_{22}-a_{12}a_{21})$$

$$q_1=c_1b_1+c_2b_2-d(a_{11}+a_{22})$$

$$q_2=d. \qquad ...(4)$$

Left hand side of the differential equation is the same for all system variables and the only difference between any of the differential equations describing any system variable is in the constant coefficients q_2, q_1 and q_0 on the right hand side.

Impulse Response of First-order RC Circuits

The series RC circuit in the below figure (a) has zero initial condition and it is excited by a unit impulse voltage source. The capacitor cannot absorb the impulse voltage. Hence, the resistor absorbs the impulse voltage and as a result, an impulse current of 1/RC flows through the circuit.

This impulse current flow results in sudden dumping of 1/RC of charge on the capacitor plates. Thus, the capacitor voltage is changed from 0 at $t = 0^-$ to 1/RC at $t = 0^+$.

The unit impulse voltage source is a short circuit for $t \geq 0^+$. Therefore, the only effect of impulse voltage application is to change the initial condition of capacitor instantaneously. The circuit becomes a source-five circuit with initial energy for $t \geq 0^+$ and executes its zero-input response. The relevant circuit is shown in the below figure (b).

Initial voltage across capacitor is 1/RC and all the voltages and currents in the circuit decay exponentially to zero with a time constant of $\tau = RC$.

$$v_C(t) = -v_R(t) = \frac{1}{RC} e^{-t/\tau} \text{ V for } t \geq 0^+$$

$$-i_C(t) = -i_R(t) = \frac{1}{R^2C} e^{-t/\tau} \text{ A for } t \geq 0^+$$

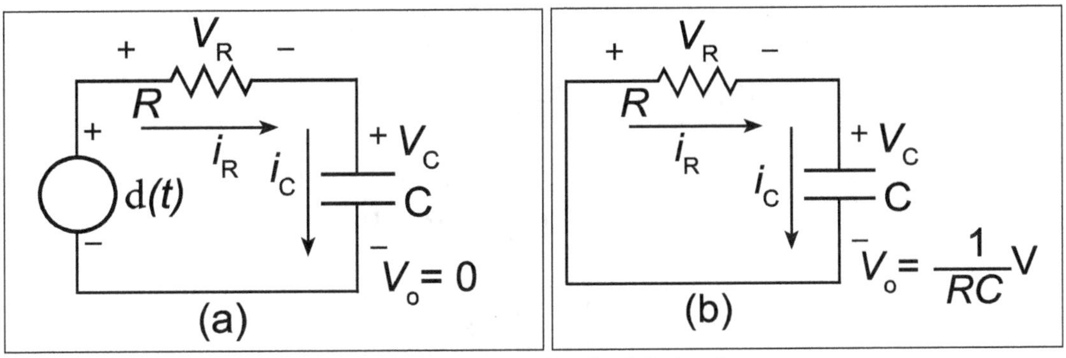

Impulse response of series RC circuit.

A capacitor with an initial voltage of V_0 across it at $t = 0$ is replaced by a capacitor with zero initial voltage and an impulse current source of suitable magnitude (CV_0) and suitable polarity is connected across it.

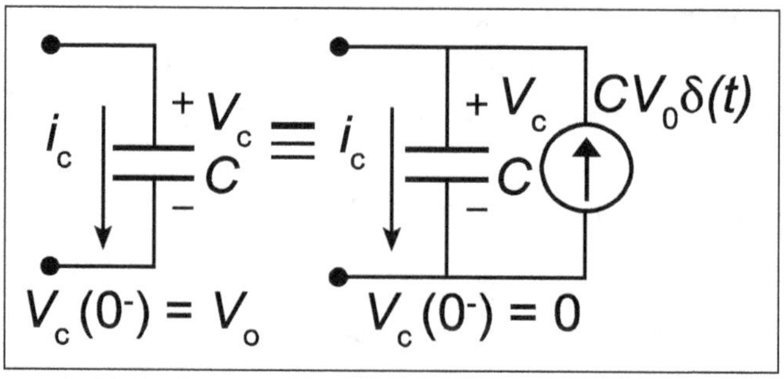

Non-zero initial voltage and impulse current application in RC circuit.

The below figure shows the application of an unit impulse current to a parallel RC circuit with zero initial energy. The resistor cannot support the impulse current.

If it were to do so, it would have called an impulse voltage across it and it will be resisted by the capacitor in parallel. Therefore, all the impulse content goes through the capacitor where the voltage is changed to 1/C from 0 at t = 0 to 1/C V at t = 0^+.

The unit impulse current source is an open-circuit after t = 0^+. Therefore, the circuit is a source-free circuit for t ≥ 0^+ and executes its zero-input response as shown in the below figure (b).

Therefore,

$$v_C(t) = v_R(t) = \frac{1}{C} e^{-t/\tau} \text{ V for } t \geq 0^+$$

$$i_R(t) = -i_C(t) = \frac{1}{RC} e^{-t/\tau} \text{ A for } t \geq 0^+$$

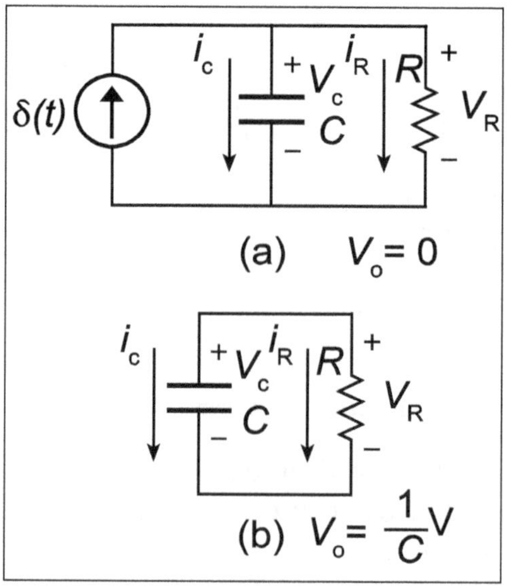

Unit impulse response of parallel RC circuit.

Problems

The value of I in the below circuit is 3 x 10^{-6} C. Let us find the zero-state response for the current through the 10kΩ resistor.

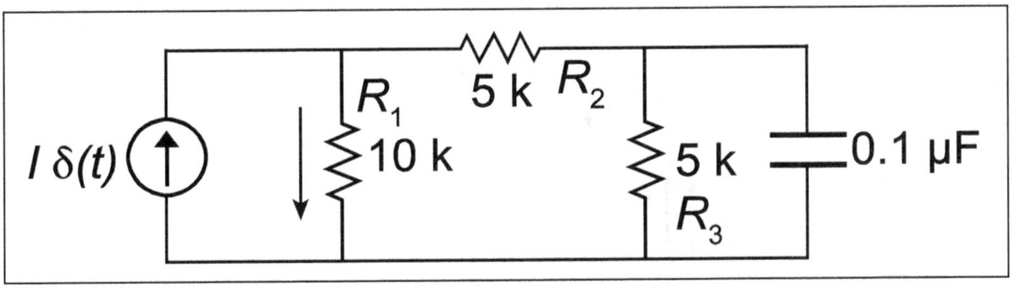

Solution:

The voltage across the capacitor changes by a finite amount as a result of impulse current flow. Current through the 5kΩ resistor (R_3) across the capacitor cannot be an impulse. Therefore, the capacitor shorts the 5 kΩ resistor across it as far as the impulse current flow is concerned. Hence, 3 x 10^{-6} δ(t) is shorted by 10kΩ and the other 5 kΩ (R_2) is shorted as per the current division principle in parallel resistors.

Thus, 2 x 10^{-6} δ(t) goes through the other 5 kΩ resistor (R_2) and the 0.1 μF capacitor. This results in sudden dumping of 2 μC of charge across the capacitor which raises the voltage to 20 V at t = 0^+. The current source is open for t > 0^+ and the circuit executes its zero-input response.

Time constant of the circuit is found by connecting the equivalent resistance across the capacitor. Since τ = 0.375 ms, the discharge current from the capacitor at 0- is 20 V/3.75 kΩ = 5.333 mA and 20 V/15 kΩ = 1.333 mA which goes through the 10 kΩ in the direction marked at that instant. Therefore, the zero-state response of this current for Iδ(t) current excitation at input of the circuit is $1.333e^{-2666.67t}$ mA.

Unit Impulse Response of Series RL Circuit

$$v_S(t) = \delta(t)$$

where,

δ (t) = Unit impulse voltage.

Unit impulse function is a limiting case of a rectangular pulse waveform of amplitude 1/ Δt V which is located between t = 0 and t = Δt as Δt ≥ 0.

The pulse always maintains unit area under it by increasing the amplitude as its duration decreases. Unit impulse function is used to analyses impulse response of RL circuit where the circuit is assumed to be initially relaxed.

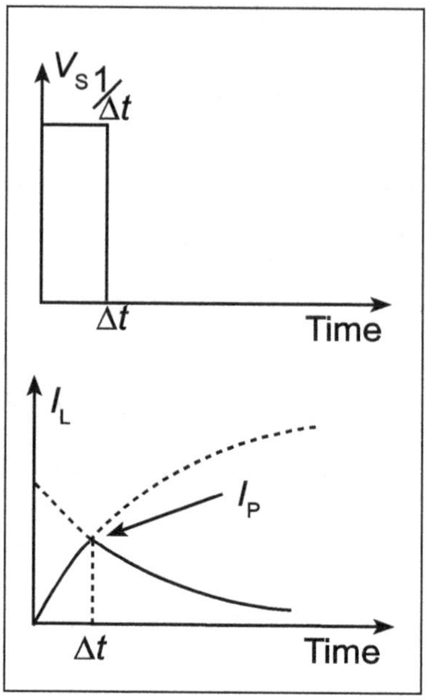

Pulse response of series RL circuit.

The applied pulse and circuit current response are shown in the above figure. The response in the interval $[0^+ \Delta t]$ is same as the unit step response in initially relaxed RL circuit which is scaled by $1/\Delta t$.

At Δt, the applied voltage at input is zero i.e., the input is shorted and a source-free RL circuit executes its zero-input response from that point onwards. An expression for I_p is obtained by evaluating the first part of the response at Δt as shown in the below equation (1).

$$i_L(t) = \begin{cases} \dfrac{1}{R\Delta t}\left(1 - e^{-t/\tau}\right) \text{for } 0^+ \le t \le \Delta t \\ I_P\, e^{-(t-\Delta t)/\tau} \text{ for } \Delta t < t < \infty \end{cases} \quad \text{...(1)}$$

$$i_P = \frac{1}{R\Delta t}\left(1 - e^{-\Delta t/\tau}\right)$$

As Δt is decreased, the value of I_p increases and moves towards left in time axis. It attains a limit as $\Delta t \to 0$. This limiting value is found by using series expansion for exponential function as in the below equation (2).

$$i_P = \frac{1}{R\Delta t}\left(1 - e^{-\Delta t/\tau}\right)$$

$$= \frac{1}{R\Delta t}\left[1 - \left\{1 - \frac{\Delta t}{\tau} + \frac{1}{2!}\left[\frac{\Delta t}{\tau}\right]^2 - \frac{1}{3!}\left[\frac{\Delta t}{\tau}\right]^3 + \ldots\right\}\right] \quad \text{...(2)}$$

$$=\frac{1}{R\tau} \text{for small} \frac{\Delta t}{\tau}$$

$$=\frac{1}{L} (\because \tau = L/R)$$

When $\Delta t \rightarrow 0$, the first part of the response jumps by 1/L.

The second part becomes an exponential from $t = 0^+$ onwards with 1/L as the starting value.

The impulse response is given as:

$$I_L(t) = \frac{1}{L} e^{-t/\tau} \text{ A for } t \geq 0^+. \qquad ...(3)$$

Only response component that can be there in the impulse response is the transient response component. Since the initial condition at $t = 0^-$ is zero, it follows that zero-input response is zero. The impulse response appearing in the above equation (3) is the zero-state response for an unit impulse application.

Let us consider the series RL circuit with an initial condition of 1/L at $t = 0^-$ and $v_s(t) = 0$ for $t \geq 0^+$. This is a source-free RL circuit with non-zero initial energy. It will have only zero-input response component in its response and its total response is given as:

$$i_L(t) = \frac{1}{L} e^{-t/\tau} \text{ A for } t \geq 0^+$$

It is same as the zero-state unit impulse response. The effect of applying δ (t) is a change in the initial condition of the inductor between $t = 0^-$ and $t = 0^+$.

Initial condition at $t = 0^-$ and $t = 0^+$ is same only if there is no impulse voltage in the circuit. δ(t) keeps the circuit shorted for $t \leq 0^-$ and $t \geq 0^+$ and an undefined voltage is applied at t = 0 where a finite volt-second area of 1 volt-sec is dumped into the circuit at t = 0. This changes the inductor current between $t = 0^-$ and $t = 0^+$.

$$i_L(t) = \frac{1}{L} \int_{-\infty}^{I} v_L(t) dt$$

$$= \underbrace{\frac{1}{L} \int_{-\infty}^{0^-} v_L(t) dt + \frac{1}{L} \int_{0^-}^{0^+} v_L(t) dt}_{=I} \frac{1}{L} \int_{0^+}^{I} v_L(t) dt \qquad ...(4)$$

The first integral gives the initial current in the inductor due to all the voltages applied to it in the past. The second integral will be non-zero only if the impulse voltage is applied to the inductor at t = 0.

As far as $I_L(t)$ after $t = 0^+$ is concerned, these two terms are interchangeable. Therefore, an initial current of I_0 in the inductor at $t = 0^-$ is replaced by zero initial current at $t = 0^-$ and an impulse voltage $LI_0\delta(t)$ has correct polarity in series with the inductor.

Voltage variable that appeared in the above equation (4) was the voltage across the inductor. If we want to replace a non-zero initial current at $t = 0^-$ by an impulse voltage source, we must ensure that the impulse source appears fully across the inductor and it does not lose itself across other elements in the circuit.

The definition of an unit impulse function avoids defining it at t = 0 and makes up for that by providing its area content in an infinitesimally small interval around t = 0.

$$\delta(t)=\begin{cases} 0 \text{ for } -\infty < t \le 0^- \\ \text{undefined at } t = 0 \\ 0 \text{ for } 0^+ \le t < \infty \end{cases}$$

$$\int_{0^-}^{0^+} \delta(t)\, dt = 1 \qquad ...(5)$$

If we apply KVL in the mesh, we get:

$$v_R(t) + v_L(t) = v_S(t) \text{ for all } t$$

Since this relation is true for all t, we can integrate both sides of the equation between the same two limits to get the following relation.

Area under $v_R(t)$ between two instants t_1 and t_2 and area under $v_L(t)$ between two instants t_1 and t_2 = Area under $v_S(t)$ between two instants t_1 and t_2

At $t_1 = 0^-$ and $t_2 = 0^+$, we have:

$$\int_{0^-}^{0^+} v_R(t)dt + \int_{0^-}^{0^+} v_L(t)dt = \int_{0^-}^{0^+} \delta(t)dt = 1 \qquad ...(6)$$

First integral on the left side is zero. Let us assume that it is zero. Then second integral is 1. The second integral value divided by L gives the change in inductor current over the interval $[0^-, 0^+]$. Therefore, the inductor current change is 1/L since the circuit is initially relaxed.

This is the maximum value that the current can have at $t = 0^+$ where integral value is zero. Thus, the current through the circuit during $[0^-, 0^+]$ is confined between 0 and a maximum of $1/L$. This implies that though the current during this infinitesimal interval is indeterminate in value, it remains upper-bounded and lower-bounded which is finite-valued in the interval.

A finite-valued function integrated over infinitesimal interval results in zero integral value. There is no area under a rectangle if the rectangle is of finite length in one direction and of infinitesimal length in the other direction.

The first integral i.e., the portion of area content of $\delta(t)$ lost across resistor is zero. This integral is non-zero and the assumption leads to a contradiction. Therefore, all the area content of $\delta(t)$ appears across the inductor itself.

Therefore, we can assert that a non-zero initial current of I_0 in an inductor L at $t = 0^-$ in a circuit is replaced by zero initial current at $t = 0^-$ along with an impulse voltage source $LI_0\ \delta(t)$ in series with the inductor for solving the circuit in the domain $[0^+, \infty)$.

Each initial condition represents a source and it is a multi-source problem. When changes are effected in a source, the superposition principle has to be applied to that component of the total response contributed by the particular source.

Series RC Circuit with Real Exponential Input

Let us consider a series RC circuit which is excited at the input by a real exponential voltage source of the form $v_s(t) = e^{-\alpha t}\, u(t)$ V. Under real exponential excitation in the series RL circuit, the response is at zero-state.

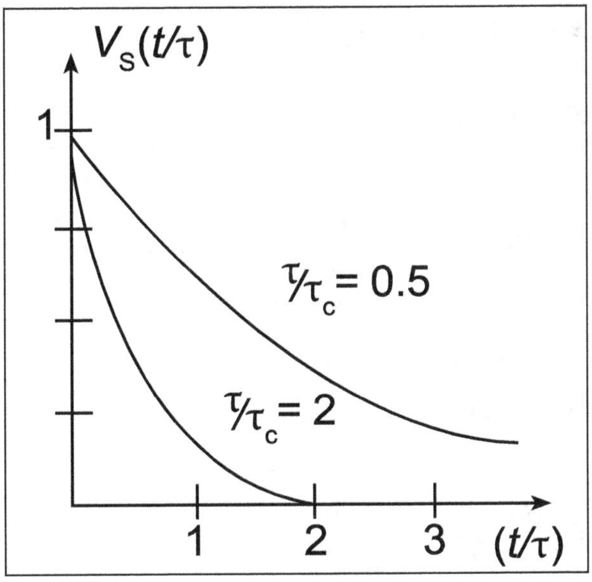

(a) Input wave.

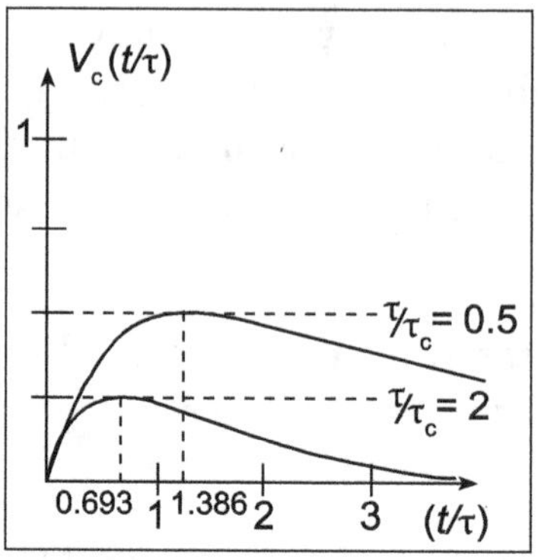

(b) Capacitor voltage.
Zero state response of RC circuit for an exponential input.

The zero-state response of the capacitor voltage is given as:

$$v_C(t)=\frac{\alpha}{\alpha-\sigma}\left(e^{-\alpha l}-e^{-\alpha l}\right)V \text{ for } t\geq o^+, \text{where } \alpha = \frac{1}{RC} \text{ and } \alpha \neq \sigma$$

$$v_C(t)=\left(\frac{1}{1-\frac{\tau}{\tau_c}}\right)\left(e^{-t\tau/\tau_4}-e^{-1}\right)V \text{ for } t\geq o^+,$$

where $\tau=RC, \tau_c = \frac{1}{\sigma}, t_n = \frac{t}{\tau}$ and $\tau\neq\tau_c$. ...(2)

The input source functions and the corresponding capacitor voltage waveforms are shown in the above figure. Time is normalized to the base of circuit time constant as shown in the above figure.

The case with $\alpha = \sigma$ was avoided in the earlier assumptions.

$$v_C(t)=\left(\frac{\alpha}{\alpha-\sigma}\right)\left(e^{-\sigma 1}-e^{-\sigma 1}\right) \text{ for } t\geq o^+$$

$$=\left(\frac{\alpha}{\alpha-\sigma}\right)\left[\left(1-\sigma 1+\frac{(\sigma t)^2}{2!}-\frac{(\sigma t)^3}{3!}+\frac{(\sigma t)^4}{4!}\ldots+\right)-\left(1-\alpha 1+\frac{(\alpha t)^2}{2!}-\frac{(\alpha t)^3}{3!}+\frac{(\alpha t)^4}{4!}\ldots+\right)\right]$$

$$=\left(\frac{\alpha}{\alpha-\sigma}\right)\left[(\alpha-\sigma)t-\frac{(\alpha^2-\sigma^2)}{2!}t^2+\frac{(\alpha^3-\sigma^3)}{3!}t^3-\frac{(\alpha^4-\sigma^4)}{4!}t^4+\ldots\right]$$

$$=(\alpha t)\left[1+\sum_{t=2}^{\infty}(-1)^{i-1}\frac{(\alpha^i-\sigma^i)}{(\alpha-\sigma)i!}t^{i-1}\right]$$

$$\left(\alpha^{i}-\sigma^{i}\right)=(\alpha-\sigma)\underbrace{\left(\alpha^{i-1}+\alpha^{i-2}\sigma+\alpha^{i-3}\sigma^{2}+\ldots.+\alpha\sigma^{I-2}+\sigma^{i-1}\right)}_{\text{i terms}}$$

$$\therefore \text{ As } \alpha\to\sigma, \frac{\left(\alpha^{i}-\sigma^{i}\right)}{(\alpha-\sigma)}\to i\times\alpha^{i-1} \text{ and } v_C(t)\to(\alpha t)\left[1+\sum_{i=2}^{\infty}(-1)^{i-1}\frac{i\times\alpha^{i-1}}{i!}t^{i-1}\right]$$

$$\text{i.e., } v_C(t)\to(\alpha t)\left[1-\alpha t+\frac{(\alpha t)^{2}}{2!}-\frac{(\alpha t)^{3}}{3!}+\frac{(\alpha t)^{4}}{4!}+\ldots\right]$$

$$\therefore v_C(t)=\alpha t e^{-\alpha t}=\left(\frac{t}{\tau}\right)e^{-\left(\frac{t}{\tau}\right)} \text{V for} \geq 0^{+}$$

The wave-shape of the output voltage is same as in the above figure. The output in this case will reach a maximum of I/e = 0.3678 at t = τ.

Problem

Two first-order series RC circuit are cascaded using a unity gain buffer amplifier as shown in the below figure. Let us find the output $v_0(t)$ as a function of time if the circuit is initially relaxed.

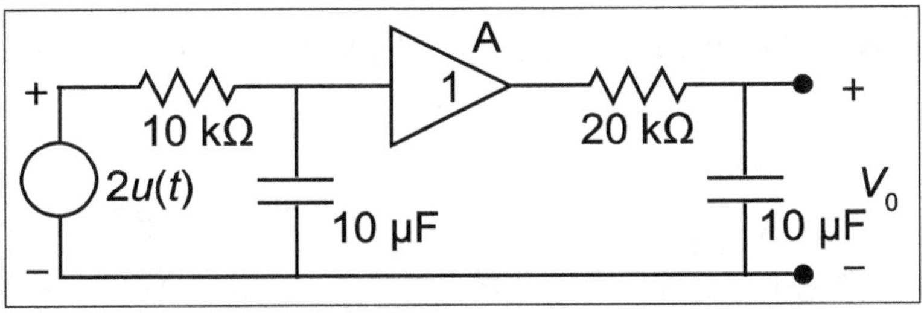

Solution:

The unity gain buffer amplifier prevents any loading-interaction between the two RC circuits. This implies that the response of the first RC stage is independent of the presence of the second stage. The first stage produces a voltage across its capacitor which is accepted by second stage as its input source function comes from an ideal independent voltage source.

Let $v_1(t)$ be the response voltage of the terminals of the first capacitor. Then, $v_1(t)$ is twice the zero-state response to unit step input (i.e., step response).

Time constant of first stage is given as:

$$10\ \text{k}\Omega\times10\ \mu\text{F}=100\ \text{ms}=0.1\ \text{s}$$

$$V_1(t) = 2\left(1 - e^{-10t}\right) \text{ V for f} \geq 0^+$$

This voltage is the input to second stage since the gain of the buffer amplifier is unity. This input is treated as the sum of two inputs namely 2u(t) and -2^{e-10t} u(t).

Zero-state response of a lumped linear time-invariant circuit obeys superposition principle and the responses to these inputs is found individually and they are superposed to get the desired response.

The time constant of second stage is given as:

$$20 \text{ k}\Omega \times 10 \text{ }\mu\text{F} = 200 \text{ ms} - 0.2 \text{ s}$$

The component contributed to $v_0(t)$ by 2u(t) is $2(1 - e^{-5t})$ V.

The contribution to $v_0(t)$ by $-2e^{-10t}$ u(t) is obtained by using the above equation(1) with $\alpha = 5$ and $\sigma = 10$. This contribution is $-2 \times -(e^{-10t} - e^{-5t})$ V for $t \geq 0+$.

Therefore:

$$v_0(t) = 2\left(1 - e^{5t}\right) + 2\left(e^{-10t} - e^{-5t}\right) \text{ V for } t \geq 0^+$$

$$= 2 - 4e^{-5t} + 2e^{-10t} \text{V for } t \geq 0^+$$

The variation of capacitor voltages $v_1(t)$ and $v_2(t)$ as functions of time is shown in the below figure.

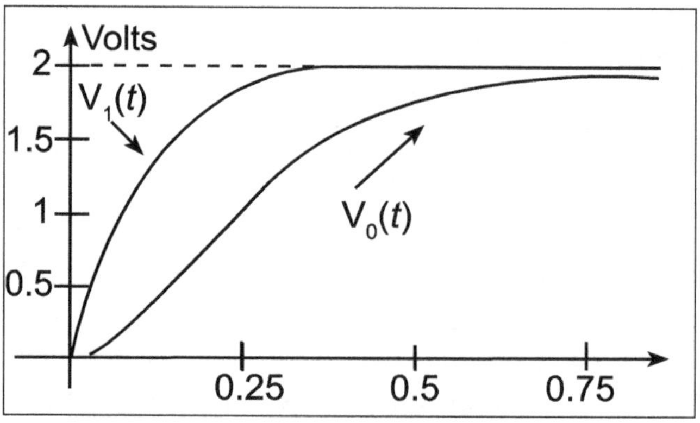

Output waveforms across RC stage capacitors.

Series RL Circuit with Exponential Inputs

Let us consider a zero-state response of series RL circuit for exponential inputs and sinusoidal inputs. Zero-input response is Io $e^{-t/\tau}$ where Io is the initial current specified at t = 0-. The total response is found by adding zero-input response and zero-state response together.

We permit exponential inputs of the form $e^{st}u(t)$ where s = - σ + jω is a complex number(σ and ω are two real numbers). A negative sign is placed behind σ since we want the real part of s to be negative when a is positive. This signal is a complex signal with a real part function and an imaginary part function as shown in the below equation (1).

$$v_s(t)=e^{st}u(t)=e^{(-\sigma+j\omega)t}u(t)=\left(e^{-\sigma t}\cos\omega t\right)u(t)+j\left(e^{-\sigma t}\sin\omega t\right)u(t) \qquad ...(1)$$

We use Euler's identity to express the complex exponential function in the rectangular form involving trigonometric functions.

Therefore, this signal does not represent a physical signal and it cannot be applied to any circuit but it can be the forcing function in a differential equation. Its real part and the imaginary part are real functions which can be created physically.

Zero-state response of a linear circuit obeys the superposition principle. Let us apply a signal x (t) to the circuit and let the zero-state response be is (t). Similarly, let the zero-state response be $i_y(t)$ when a signal y(t) is applied to the same circuit.

Let us find the response when $j_y(t)$ is applied where $j=\sqrt{-1}$.

By superposition principle, it will be $ji_y(t)$. When we apply x(t) + jy(t) as input, then we must get $i_x(t) + ji_y(t)$ as the response. The real part of the zero-state response for a complex signal input is the zero-state response for the real part of the input and the imaginary part of the zero-state response for a complex signal input is the zero -state response for the imaginary part of the input. This is a direct result of the linearity of the circuit and will not be true in non-linear circuit.

Let us obtain the zero-state response of circuits to sinusoidal inputs by taking the real or the imaginary part of the zero-state response for complex exponential inputs. Complex exponential functions will sinusoid when solving differential equations. Let us solve the zero-state response of series RL circuit with a complex exponential input.

$$\frac{di_L(t)}{d(dt)}+\frac{i_L(t)}{\tau}=\frac{1}{L}e^{st}u(t) \qquad ...(2)$$

This equation is true for all t and in particular, for all t ≥ 0^+. This is possible only if the shape of the functions on both sides of the equation are same. This will imply that the trial solution for the particular integral has to be Ae^{st}. Substituting this trial solution in the differential equation in the above equation (2), we have:

$$Ae^{st}(s+1/\tau)=(1/L)e^{st} \text{ for } t\geq 0^+$$

$$\therefore A=\frac{1}{sL+L/\tau}=\frac{1}{sL+R}\left(\because \tau=L/R\right) \qquad ...(3)$$

$$\therefore i_L(t) = \frac{1}{sL+R} e^{st} \text{ for } t \geq o^+$$

The total solution is obtained by adding this particular integral to the complementary solution. Zero-state response is found by finding a particular integral and a suitable complementary function because zero-state response involves both forced response and natural response.

$$\therefore i_L(t) = Be^{-t/\tau} + \frac{1}{sL+R} e^{st} \text{ for } t \geq o^+$$

Substituting the initial current value at o⁺, we have:

$$o = B + \frac{1}{sL+R}$$

$$\therefore B = -\frac{1}{sL+R} \text{ and} \quad \text{...(4)}$$

$$i_L(t) = \frac{1}{sL+R}\left(e^{st} - e^{-t/\tau}\right) \text{ for } t \geq o^+$$

4.3 Complete Response of RC, RL and RLC Circuits to Sinusoidal Excitation

Response of R-L-C Circuits for Sinusoidal Excitation

We know the initial conditions in the series R-L, series R-C and series R-L-C circuits. We have found out these initial conditions in various elements of network when D.C. voltage is applied. Let us determine the analysis of network with the A.C. excitation to the series circuits and let us also find the response of the series circuits for A.C. excitation.

Series R-L Circuit

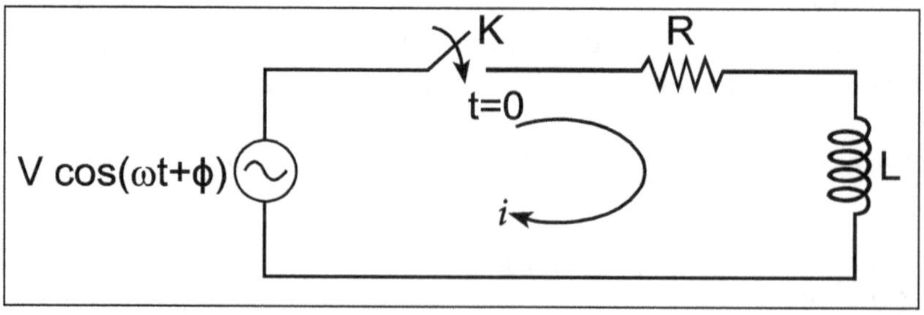

Series R-L circuit.

Let us consider the above circuit consisting of resistance and inductance in series as shown in the above figure. The switch K is closed at t = 0. At t = 0, sinusoidal voltage V cos (ωt + φ) is applied to the series R-L circuit where V is amplitude of the wave and φ is the phase angle.

After the switching has taken place, Kirchhoff's voltage law is applied to the circuit.

$$V\cos(\omega t+\phi)=Ri+L\frac{di}{dt} \qquad ...(1)$$

$$\frac{di}{dt}+\frac{R}{L}i=\frac{V}{L}\cos(\omega t+\phi)$$

For the above equation, the solution consists of two parts namely complementary function and particular integral.

The complementary function of the solution 'i' is given as:

$$i_{CF}=K_1 e^{-t(R/L)}$$

The particular solution is obtained by using undetermined coefficients.

$$i_{PI}=A\cos(\omega t+\phi)+B\sin(\omega t+\phi) \qquad ...(2)$$

$$i'_{PI}=-A\omega\sin(\omega t+\phi)+B\omega\cos(\omega t+\phi) \qquad ...(3)$$

Substituting equation (2) and (3) in the above equation (1), we have:

$$\left[-A\omega\sin(\omega t+\phi)+B\omega\cos(\omega t+\phi)\right]+\frac{R}{L}\left[A\cos(\omega t+\phi)+B\sin(\omega t+\phi)\right]=\frac{V}{L}\cos(\omega t+\phi)$$

$$\therefore\left(-A\omega+\frac{BR}{L}\right)\sin(\omega t+\phi)+\left(B\omega+\frac{AR}{L}\right)\cos(\omega t+\phi)=\frac{V}{L}\cos(\omega t+\phi)$$

By comparing sine and cosine terms, we get:

$$-A\omega+\frac{BR}{L}=0$$

$$B\omega+\frac{AR}{L}=\frac{V}{L}$$

From the above equations, we have:

$$A=V\frac{R}{R^2+(\omega L)^2}$$

$$B = V\frac{\omega L}{R^2 + (\omega L)^2}$$

Substituting the values of A and B in the above equation (2), we have:

$$i_{PI} = V\frac{R}{R^2 + (\omega L)^2}\cos(\omega t + \phi) + V\frac{\omega L}{R^2 + (\omega L)^2}\sin(\omega t + \phi) \quad ...(4)$$

$$K_2 \cos\phi = \frac{VR}{R^2 + (\omega L)^2}$$

$$K_2 \sin\phi = \frac{V\omega L}{R^2 + (\omega L)^2}$$

To find k_2 and φ, divide k_2 sin φ by cos φ, we have:

$$\tan\phi = \frac{\omega L}{R}$$

$$\phi = \tan^{-1}\frac{\omega L}{R}$$

Squaring k_2 sin φ and k_2 cos φ and adding, we have:

$$K_2^2 \sin^2\phi + K_2^2 \cos^2\phi = \frac{V^2}{R^2 + (\omega L)^2}$$

$$\therefore K_2^2 = \frac{V^2}{R^2 + (\omega L)^2}$$

$$\therefore K_2 = \frac{V}{\sqrt{R^2 + (\omega L)^2}}$$

The particular solution is given as:

$$i_{PI} = \frac{V}{\sqrt{R^2 + (\omega L)^2}}\cos\left(\omega t + \phi - \tan^{-1}\frac{\omega L}{R}\right)$$

Complete solution for the current 'i' is the addition of complementary function and particular integral.

$$\therefore\ i = i_{CF} + i_{PI}$$

$$\therefore\ i=K_1 e^{-t(R/L)}+\frac{V}{\sqrt{R^2+(\omega L)^2}}\cos\left(\omega t+\phi-\tan^{-1}\frac{\omega L}{R}\right)\ \dots 5 \qquad \dots(5)$$

To find the value of arbitrary constant K_1, we have to use initial condition. Since the inductor does not allow sudden changes at t = 0, i = 0.

Substituting this value, we have:

$$K_1=-\frac{V}{\sqrt{R^2+(\omega L)^2}}\cos\left(\phi-\tan^{-1}\frac{\omega L}{R}\right)$$

The complete solution for the current is given as:

$$i=e^{-t(R/L)}\left[\frac{-V}{\sqrt{R^2+(\omega L)^2}}\cos\left(\phi-\tan^{-1}\frac{\omega L}{R}\right)\right]$$

$$+\frac{V}{\sqrt{R^2+(\omega L)^2}}\cos\left(\omega t+\phi-\tan^{-1}\frac{\omega L}{R}\right) \qquad \dots(6)$$

Series R-C Circuit

Let us consider the below circuit which has a resistance and capacitance in series as shown in the below figure. The switch K is closed at t = 0. At t = 0, a sinusoidal voltage V cos (ωt + φ) is applied to R-C circuit where V is the amplitude of the wave and φ is the phase angle. After switching has taken place, Kirchhoff's voltage law is applied to the loop in the circuit.

$$V\cos(\omega t+\phi)=Ri+\frac{1}{C}\int i\,dt$$

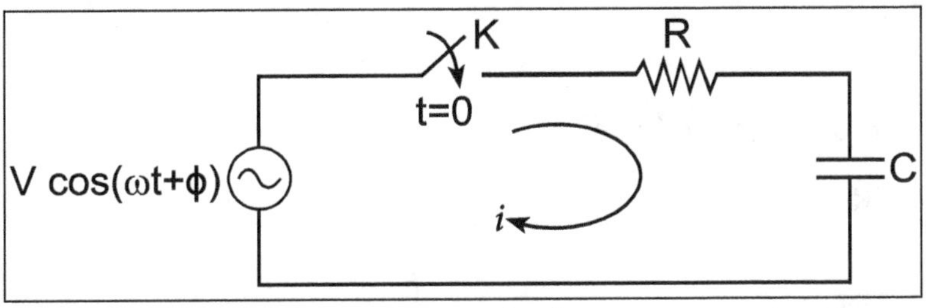

Series R-C circuit.

Differentiating the above equation with respect to t, we have:

$$\therefore\ R\frac{di}{dt}+\frac{1}{C}i=-V\omega\sin(\omega t+\phi)$$

$$\therefore \frac{di}{dt}+\frac{1}{RC}i=\frac{-V\omega}{R}\sin(\omega t+\phi) \quad ...(7)$$

For the above equation, the solution consists of two parts namely complementary function and particular integral.

Complementary function of the solution 'i' is given as:

$$i_{CF}=K_1 e^{-t/RC}$$

The particular solution can be obtained by using undetermined coefficients.

$$i_{PI}=A\cos(\omega t+\phi)+B\sin(\omega t+\phi) \quad ...(8)$$

$$i'_{PI}=-A\omega\sin(\omega t+\phi)+B\omega\cos(\omega t+\phi) \quad ...(9)$$

Substituting equation (8) and (9) in equation (7), we have:

$$\therefore\left[-A\omega\sin(\omega t+\phi)+B\omega\cos(\omega t+\phi)\right]+\frac{1}{RC}\left[A\cos(\omega t+\phi)+B\sin(\omega t+\phi)\right]$$

$$=-\frac{V\omega}{R}\sin(\omega t+\phi)$$

$$\therefore\left(-A\omega+\frac{B}{RC}\right)\sin(\omega t+\phi)+\left(B\omega+\frac{A}{RC}\right)\cos(\omega t+\phi)=\frac{-V\omega}{R}\sin(\omega t+\phi)$$

By comparing sine and cosine terms, we have:

$$-A\omega+\frac{B}{RC}=-\frac{V\omega}{R} \text{ and}$$

$$B\omega+\frac{A}{RC}=0$$

From the above equations, we have:

$$A=\frac{VR}{R^2+\left(\frac{1}{\omega C}\right)^2}$$

$$B=\frac{-V}{\omega C\left[R^2+\left(\frac{1}{\omega C}\right)^2\right]}$$

Substituting the values of A and B in equation (8), we have:

$$i_{PI} = \frac{VR}{R^2 + \left(\frac{1}{\omega C}\right)^2}\cos(\omega t + \phi) + \frac{-V}{\omega C\left[R^2 + \left(\frac{1}{\omega C}\right)^2\right]}\sin(\omega t + \phi) \quad \text{...(10)}$$

$$K_2 \cos\phi = \frac{VR}{\left[R^2 + \left(\frac{1}{\omega C}\right)^2\right]}$$

$$K_2 \sin\phi = \frac{V}{\omega C\left[R^2 + \left(\frac{1}{\omega C}\right)^2\right]}$$

To find k_2 and φ, let us divide $k_2 \sin\varphi$ by $k_2 \cos\varphi$.

$$\tan\phi = \frac{1}{\omega RC}$$

$$\phi = \tan^{-1}\left(\frac{1}{\omega RC}\right)$$

Squaring $k_2 \sin\varphi$ and $k_2 \cos\varphi$ and adding, we have:

$$K_2^2 \sin^2\phi + K_2^2 \cos^2\phi = \frac{V^2}{\left[R^2 + \left(\frac{1}{\omega C}\right)^2\right]}$$

$$K_2^2 = \frac{V^2}{\left[R^2 + \left(\frac{1}{\omega C}\right)^2\right]}$$

$$K_2 = \frac{V}{\sqrt{R^2 + \left(\frac{1}{\omega C}\right)^2}}$$

The particular solution is given as:

$$i_{PI} = \frac{V}{\sqrt{R^2 + \left(\frac{1}{\omega C}\right)^2}}\cos\left(\omega t + \phi + \tan^{-1}\frac{1}{\omega RC}\right)$$

Complete solution for the current 'i' is the addition of complementary function and particular integral.

$$i = i_{CF} + i_{PI}$$

$$i = K_1 e^{-t/RC} + \frac{V}{\sqrt{R^2 + \left(\frac{1}{\omega C}\right)^2}} \cos\left(\omega t + \phi + \tan^{-1}\frac{1}{\omega RC}\right) \quad ...(11)$$

To find value of arbitrary constant K_1, we have to use initial conditions since capacitor does not allow sudden changes in voltage at t = 0 where $i = \frac{V}{R}\cos\phi$.

Substituting these values, we have:

$$\frac{V}{R}\cos\phi = K_1 + \frac{V}{\sqrt{R^2 + \left(\frac{1}{\omega C}\right)^2}} \cos\left(\phi + \tan^{-1}\frac{1}{\omega RC}\right)$$

$$K_1 = \frac{V}{R}\cos\phi - \frac{V}{\sqrt{R^2 + \left(\frac{1}{\omega C}\right)^2}} \cos\left(\phi + \tan^{-1}\frac{1}{\omega RC}\right)$$

The complete solution for the current is given as:

$$i = e^{-t/RC}\left[\frac{V}{R}\cos\phi - \frac{V}{\sqrt{R^2 + \left(\frac{1}{\omega C}\right)^2}} \cos\left(\phi + \tan^{-1}\frac{1}{\omega RC}\right)\right]$$

$$+ \frac{V}{\sqrt{R^2 + \left(\frac{1}{\omega C}\right)^2}} \cos\left(\omega t + \phi + \tan^{-1}\frac{1}{\omega RC}\right) \quad ...(12)$$

Series R-L-C Circuit

In the below circuit, resistor, inductor and capacitor are in series. The switch K is closed at t = 0. At t = 0, sinusoidal voltage V cos ($\omega t + \varphi$) is applied to the series R-L-C circuit where V is the amplitude of the wave and φ is the phase angle.

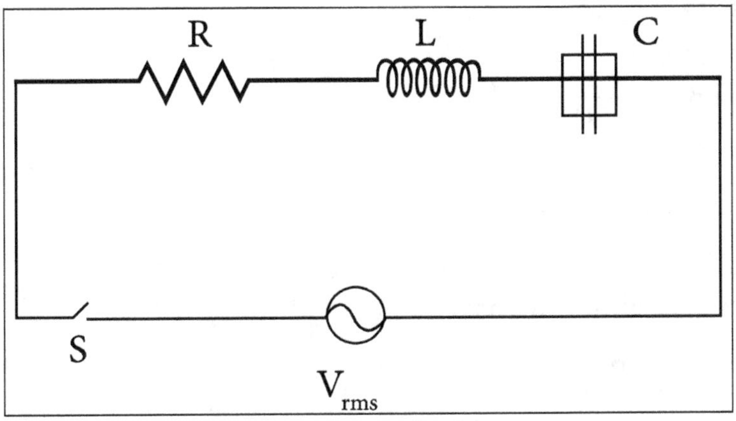

Series R-L-C circuit.

After the switching has taken place and applying Kirchhoff's voltage law to the circuit, we have:

$$V\cos(\omega t+\phi)=Ri+L\frac{di}{dt}+\frac{1}{C}\int i\,dt$$

Differentiating the above equation with respect to t, we have:

$$L\frac{d^2i}{dt^2}+R\frac{di}{dt}+\frac{1}{C}i=-V\omega\sin(\omega t+\phi)$$

$$\therefore\quad \frac{d^2i}{dt^2}+\frac{R}{L}\frac{di}{dt}+\frac{1}{LC}i=-\frac{V\omega}{L}\sin(\omega t+\phi) \qquad ...(13)$$

The above solution consists of two parts namely complementary function and particular integral.

Let us find particular solution. The particular solution is obtained by using undetermined coefficients.

$$i_{PI}=A\cos(\omega t+\phi)+B\sin(\omega t+\phi) \qquad ...(14)$$

$$i'_{PI}=-A\omega\sin(\omega t+\phi)+B\omega\cos(\omega t+\phi) \qquad ...(15)$$

$$i''_{PI}=-A\omega^2\cos(\omega t+\phi)-B\omega^2\sin(\omega t+\phi) \qquad ...(16)$$

Substituting equation (14), (15) and (16) in equation (13), we have:

$$\left[-A\omega^2\cos(\omega t+\phi)-B\omega^2\sin(\omega t+\phi)\right]+\frac{R}{L}\left[-A\omega\sin(\omega t+\phi)+B\omega\cos(\omega t+\phi)\right]+$$
$$\frac{1}{LC}\left[A\cos(\omega t+\phi)+B\sin(\omega t+\phi)\right]=\frac{-V\omega}{L}\sin(\omega t+\phi)$$

Comparing cosine and sine terms, we have:

$$A\left(\frac{\omega R}{L}\right)+B\left(\omega^2-\frac{1}{LC}\right)=\frac{V\omega}{L} \text{ and}$$

$$A\left[\omega^2-\frac{1}{LC}\right]-B\left(\frac{\omega R}{L}\right)=0$$

Solving above two equations, we have:

$$A=\frac{V\dfrac{\omega^2 R}{L^2}}{\left[\left(\dfrac{\omega R}{L}\right)^2-\left(\omega^2-\dfrac{1}{LC}\right)^2\right]} \text{ and}$$

$$B=\frac{\left(\omega^2-\dfrac{1}{LC}\right)V\omega}{L\left[\left(\dfrac{\omega R}{L}\right)^2-\left(\omega^2-\dfrac{1}{LC}\right)^2\right]}$$

$$i_{PI}=\frac{V\dfrac{w^2 R}{L^2}}{\left[\left(\dfrac{\omega R}{L}\right)^2-\left(\omega^2-\dfrac{1}{LC}\right)^2\right]}\cos(\omega t+\phi)$$

$$+\frac{\left(\omega^2-\dfrac{1}{LC}\right)V\omega}{L\left[\left(\dfrac{\omega R}{L}\right)^2-\left(\omega^2-\dfrac{1}{LC}\right)^2\right]}\sin(\omega t+\phi) \quad ...(17)$$

$$K\cos\phi=\frac{V\dfrac{w^2 R}{L^2}}{\left[\left(\dfrac{\omega R}{L}\right)^2-\left(\omega^2-\dfrac{1}{LC}\right)^2\right]} \text{ and}$$

$$K\sin\phi=\frac{\left(\omega^2-\dfrac{1}{LC}\right)V\omega}{\left[\left(\dfrac{\omega R}{L}\right)^2-\left(\omega^2-\dfrac{1}{LC}\right)^2\right]}$$

To find K and φ, divide K sin φ by K cos φ as follows.

$$\tan\phi = \frac{\left(\omega L - \frac{1}{\omega C}\right)}{R}$$

$$\phi = \tan^{-1}\left[\left(\omega L - \frac{1}{\omega C}\right)/R\right]$$

Squaring K sinφ and K cosφ and adding, we have:

$$K^2\cos^2\phi + K^2\sin^2\phi = \frac{V^2}{R^2 + \left(\frac{1}{\omega C} - \omega L\right)^2}$$

$$\therefore\ K^2 = \frac{V^2}{R^2 + \left(\frac{1}{\omega C} - \omega L\right)^2}$$

$$\therefore\ K = \frac{V}{\sqrt{R^2 + \left(\frac{1}{\omega C} - \omega L\right)^2}}$$

Particular solution is given as:

$$i_{PI} = \frac{V}{\sqrt{R^2 + \left(\frac{1}{\omega C} - \omega L\right)^2}}\cos\left[\omega t + \phi + \tan^{-1}\frac{\left(\frac{1}{\omega C} - \omega L\right)}{R}\right] \quad ...(18)$$

To find complementary function, we have characteristic equation as follows.

$$P^2 + \frac{R}{L}P + \frac{1}{LC} = 0 \quad ...(19)$$

The roots are given as:

$$P_1, P_2 = \frac{-R}{2L} \pm \sqrt{\left(\frac{R}{2L}\right)^2 - \frac{1}{LC}}$$

$$K_1 = \frac{-R}{2L} \text{ and } K_2 = \sqrt{\left(\frac{R}{2L}\right)^2 - \frac{1}{LC}}$$

$$P_1 = (K_1 + K_2)$$

$$P_2 = (K_1 - K_2)$$

Above complementary function is similar to that of D.C. series R-L-C circuit. So we will have three cases depending on nature of roots.

Case (i):

When the roots are real and unequal, the system gives over damped response. The above equation (19) becomes:

$$\left[P-(K_1+K_2)\right]\left[P-(K_1-K_2)\right]=0$$

Complementary function for the above equation is given as:

$$i_{CF} = K_3\, e^{(K_1+K_2)t} + K_4 \cdot e^{(K_1-K_2)t}$$

Complete solution is given as:

$$i = i_{CF} + i_{PI}$$

$$\therefore i = K_3\, e^{(K_1+K_2)t} + K_4 \cdot e^{(K_1-K_2)t} + \frac{V}{\sqrt{R^2 + \left(\frac{1}{\omega C} - \omega L\right)^2}} \cos\left[\omega t + \phi + \tan^{-1}\left(\frac{1}{\omega CR} - \frac{\omega L}{R}\right)\right]$$

Case (ii):

When $\frac{R^2}{4L^2} < \frac{1}{LC}$, K_2 is negative.

Then roots are complex conjugate which gives under damped system.

The above equation becomes:

$$\left[P-(K_1 + j\,K_2)\right]\left[P-(K_1 - j\,K_2)\right]=0$$

Complementary function for the above equation is given as,

$$i_{CF} = e^{K_1 t}\left[K_3 \cos K_2\, t + K_4 \sin K_2\, t\right]$$

Complete solution is given as:

$$i = i_{CF} + i_{PI}$$

$$\therefore\ i = e^{K_1 t}\left[K_3 \cos K_2 t + K_4 \sin K_2 t\right] + \frac{V}{\sqrt{R^2 + \left(\frac{1}{\omega C} - \omega L\right)^2}} \cos\left[\omega t + \phi + \tan^{-1}\left(\frac{1}{\omega CR} - \frac{\omega L}{R}\right)\right]$$

Case (iii):

When $\frac{R^2}{4L^2} = \frac{1}{LC}$, , K_2 is zero.

Then roots are equal which gives critically damped response.

The above equation becomes:

$$[P - K_1][P - K_1] = 0$$

Complementary function for the above equation is given as:

$$i_{CF} = e^{K_1 t}\left(K_3 + K_4 t\right)$$

So the complete solution is given as:

$$i = e^{K_1 t}\left(K_3 + K_4 t\right) + \frac{v}{\sqrt{R^2 + \left(\frac{1}{\omega C} - \omega L\right)}} \cos\left[\omega t + \phi + \tan^{-1}\left(\frac{1}{\omega RC} - \frac{\omega L}{R}\right)\right]$$

5

Two Port Networks

5.1 Z, Y, Transmission (ABCD) and Hybrid (H) Parameters

Consider a two port network as shown in the below figure. In all there are four variables; two voltages and two currents. In general, any two port network has one pair of voltage and current at each port.

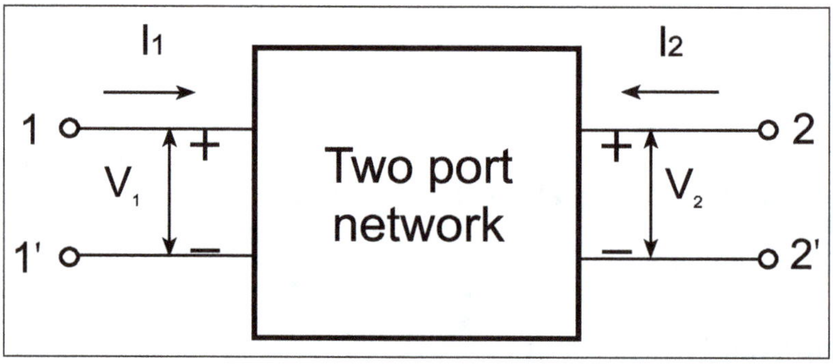

Two port network.

Assuming the variables at input and output ports as transformed quantities, the voltage and current at input terminals are V_1 and I_1 while at output terminals are V_2 and I_2, as shown in the above figure. The directions of both the currents I_1 and I_2 are assumed to be flowing into the network such currents entering the network are assumed to be positive. Also the voltage has the positive. Also the voltages have the positive reference polarities.

Before we know about the two port network parameters, let us make certain assumptions necessary in the analysis.

The assumptions are as follows:

- The voltages and currents in the network present inside a box are not available for the measurements.
- The network should consist only linear elements along with dependent sources if any. But independent or active sources should not be present in the network inside box.
- If the network inside box consists energy storing elements such as inductor and capacitor, then initial conditions of such elements should be zero.

In order to describe the relationship between port voltages and currents, one requires the linear equations equal to the number of ports. So in two port network analysis, we will require two linear equations in terms of four above mentioned variables. We can obtain these equations by considering two variables as dependent variables while other as independent variables. As the network consists only linear elements, the linear relationship can be obtained by writing two variables in terms of other two variables. There are six possible ways of selecting two independent variables out of four variables. Thus there are six different pairs of equations defining their own sets of parameters such impedance (z), admittance (y), hybrid (h), inverse hybrid (g), transmission and inverse transmission parameters.

5.1.1 Z-Parameters

These are also called impedance parameters. These are obtained by expressing voltages at two ports in terms of currents at two ports. Thus, currents I_1 and I_2 are independent variables; while V_1 and V_2 are dependent variables.

Thus, we have:

$$V_1 = f_1(I_1, I_2)$$

$$V_2 = f_2(I_1, I_2)$$

In equation form, above relations can be written as:

$$V_1 = z_{11} I_1 + z_{12} I_2 \quad ...(1)$$

$$V_2 = z_{21} I_1 + z_{22} I_2 \quad ...(2)$$

In matrix form, above equations can be written as:

$$\begin{bmatrix} V_1 \\ V_2 \end{bmatrix} = \begin{bmatrix} z_{11} & z_{12} \\ z_{21} & z_{22} \end{bmatrix} \begin{bmatrix} I_1 \\ I_2 \end{bmatrix}$$

or, $[V]=[z][I]$

The individual z-parameters can be obtained by assigning values of independent variables to be zero. The z-parameters can be defined as follows.

(i) Let $I_2 = 0$; port - 2 is open circuited.

From the above equation (1):

$$z_{11} = \left.\frac{V_1}{I_1}\right|_{I_2=0} \ \Omega$$

The parameter z_{11} is called open circuit driving point input impedance.

From equation (2):

$$z_{21} = \left.\frac{V_2}{I_1}\right|_{I_2=0} \ \Omega$$

The parameters z_{21} is called open circuit forward transfer impedance.

(ii) Let $I_1 = 0$; i.e. port - 1 is open circuited.

From equation (1):

$$z_{12} = \left.\frac{V_1}{I_2}\right|_{I_1=0} \ \Omega$$

The parameter z_{12} is called open circuit reverse transfer impedance.

From equation (2):

$$z_{22} = \left.\frac{V_2}{I_2}\right|_{I_1=0} \ \Omega$$

The parameter z_{22} is called open circuit driving point output impedance.

From above equations it is clear that z_{11} and z_{22} gives the relationship between voltages and currents at port 1 and port 2 respectively. Such parameters are the driving point impedances. While z_{12} and z_{21} gives the relationship between voltage and currents at different ports. Such parameters are the transfer impedances.

Moreover, these parameters are defined only when the current in one of the ports is zero. This corresponds to the conditions that the one of the ports is open circuited. Hence z-parameters are named as open circuit impedance parameters.

The equivalent circuit of a two port network in terms of open circuit impedance parameters satisfying equations (1) and (2) is as shown in the below figure.

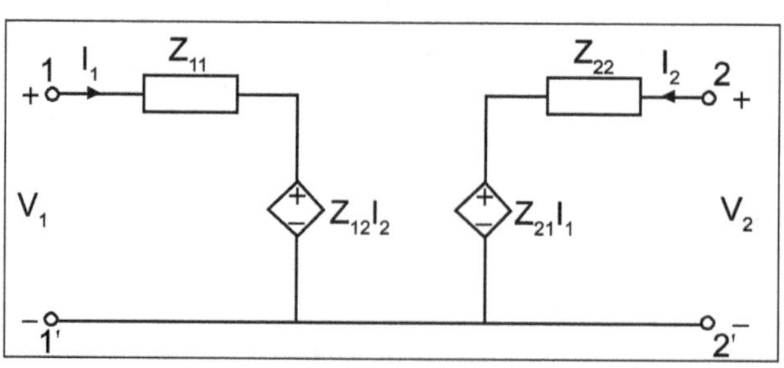

Equivalent network of a two port network in terms of z-parameters.

5.1.2 Y-Parameters

These are also called admittance parameters. These are obtained by expressing currents at two ports in terms of voltages at two ports. Thus, voltages V_1 and V_2 are independent variables, while I_1 and I_2 are dependent variables.

Thus, we have:

$$I_1 = f_1(V_1, V_2)$$

$$I_2 = f_2(V_1, V_2)$$

In equation form, above relations can be written as:

$$I_1 = y_{11} V_1 + y_{12} V_2 \quad ...(1)$$

$$I_2 = y_{21} V_1 + y_{22} V_2 \quad ...(2)$$

In matrix form, above equations can be written as:

$$\begin{bmatrix} I_1 \\ I_2 \end{bmatrix} = \begin{bmatrix} y_{11} & y_{12} \\ y_{21} & y_{22} \end{bmatrix} \begin{bmatrix} V_1 \\ V_2 \end{bmatrix}$$

$$\text{or, } [I] = [y][V]$$

The individual y-parameters are defined as follows:

(i) Let $V_2 = 0$ i.e. port-2 is short circuited.

From equation (1):

$$y_{11} = \left.\frac{I_1}{V_1}\right|_{V_2=0} \quad \mho$$

The parameter y_{11} is called short circuit driving point input admittance.

From equation (2):

$$y_{21} = \left.\frac{I_2}{V_1}\right|_{V_2=0} \quad \mho$$

The parameter y_{21} is called short circuit forward transfer admittance.

(ii) Let $V_1 = 0$ i.e. port-1 is short circuited.

From equation (1):

$$y_{12} = \left.\frac{I_1}{V_2}\right|_{V_1=0} \ \mho$$

The parameter y_{12} is called short circuit reverse transfer admittance.

From equation (2):

$$y_{22} = \left.\frac{I_2}{V_2}\right|_{V_1=0} \ \mho$$

The parameter y_{22} is called short circuit driving point output admittance.

These parameters are defined individually only when the voltage in any one of the ports is zero. This corresponds to the condition that one of the ports is short circuited. Hence y-parameters are also called short circuit admittance parameters.

The equivalent circuit of a two port network in terms of short circuit admittance parameters satisfying equations (1) and (2) is as shown in the below figure.

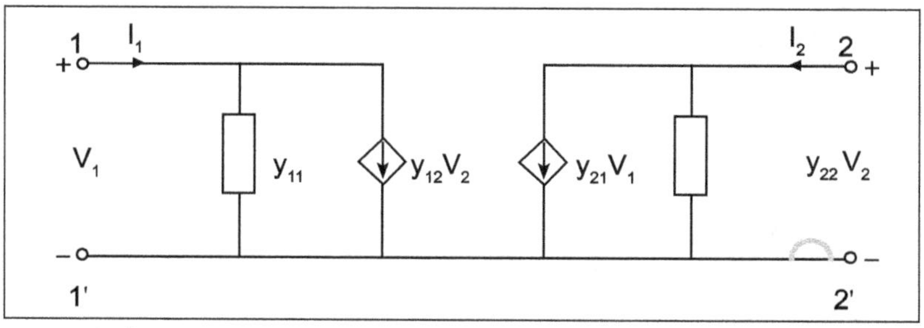

Equivalent network of a two port network in terms of y-parameters.

5.1.3 Transmission (ABCD) Parameters

These parameters are known as transmission parameters. These are generally used in the analysis of power transmission in which the input port is referred as the sending end while the output port is referred as receiving end. These are obtained by expressing voltage V_1 and current I_1 at input port in terms of voltage V_2 and current I_2 at output port. Thus, voltage V_2 and current I_2 are independent variables while voltage V_1 and current I_2 are dependent variables.

Thus, we have:

$$V_1 = f_1(V_2, -I_2)$$

$$I_1 = f_2(V_2, -I_2)$$

Generally we have considered the currents in both the ports are entering the port and both are positive. The negative sign with I_2 indicates that, for the ABCD parameters the current I_2 is leaving the port-2.

In equation form, above relations can be written as:

$$V_1 = A V_2 + B(-I_2) \quad \text{...(1)}$$

$$I_1 = C V_2 + D(-I_2) \quad \text{...(2)}$$

In matrix form, above equations can be written as:

$$\begin{bmatrix} V_1 \\ I_1 \end{bmatrix} = \begin{bmatrix} A & B \\ C & D \end{bmatrix} \begin{bmatrix} V_2 \\ -I_2 \end{bmatrix}$$

The individual ABCD parameters can be defined as follows.

(i) Let - I_2 = 0, i.e. port - 2 is circuited.

From equation (1):

$$A = \left.\frac{V_1}{V_2}\right|_{-I_2 = 0}$$

The parameter A is called open circuit reverse voltage gain. It is unitless.

From equation (2):

$$C = \left.\frac{I_1}{V_2}\right|_{-I_2 = 0} \quad \mho$$

The parameter C is called open circuit reverse transfer admittance.

(ii) Let V_2 = 0, i.e. port - 2 is short circuited.

From equation (1):

$$B = \left.\frac{V_1}{-I_2}\right|_{V_2 = 0} \quad \mho$$

The parameter B is called short circuit reverse transfer impedance.

From equation (2):

$$D = \left.\frac{I_1}{-I_2}\right|_{V_2=0}$$

The parameter D is called short circuit reverse current gain. It is also unitless.

ABCD or Transmission parameters are also called chain parameters.

These parameter are effectively used for the analysis of power transmission line, so commonly known as transmission parameters. These parameters are also useful in the analysis of the networks connected in cascade forming a chain of networks. Hence, these parameters are also called chain parameters. Let us consider a transmission line represented as a two port network and a general two port network as shown in the below figures (a) and (b) respectively.

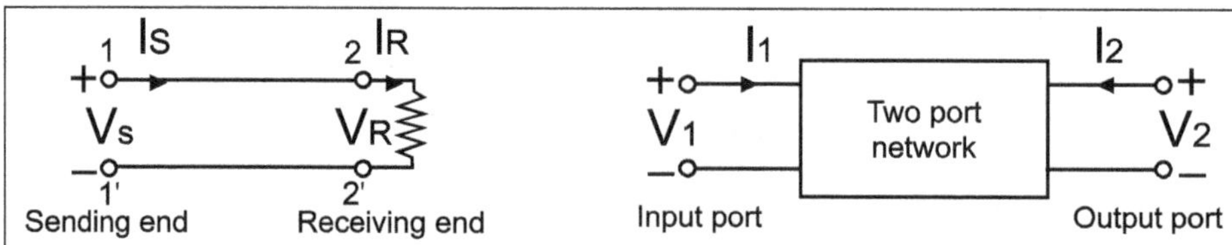

(a) Transmission line of length *l*

(b) Two port network

From above representation, it is clear that to current at port-2 flows into the network while current at the receiving end of a transmission line flows out of the network. Hence a two port network can be used in the analysis of the transmission line only if I_2 is considered to be current flowing out of terminal. Hence while considering I_2 as independent variable it is considered - I_2.

5.1.4 Hybrid (H) Parameters

BJT is a two port device in which one terminal is common to both the input and output ports. The behaviour of a two port network is analyzed using the current and voltage parameters at the input and output ports namely input current, input voltage, output current and output voltage.

Out of these four parameters, two parameters are independent parameters and the remaining two parameters are dependent parameters. The dependent parameters are expressed in terms of the independent parameters.

Let us consider the two port network shown in the below figure. The terminal behaviour of any two port network is specified by the terminal voltages v_1 and v_2 at ports 1 and 2 and currents i_1 and i_2 enters the ports 1 and 2.

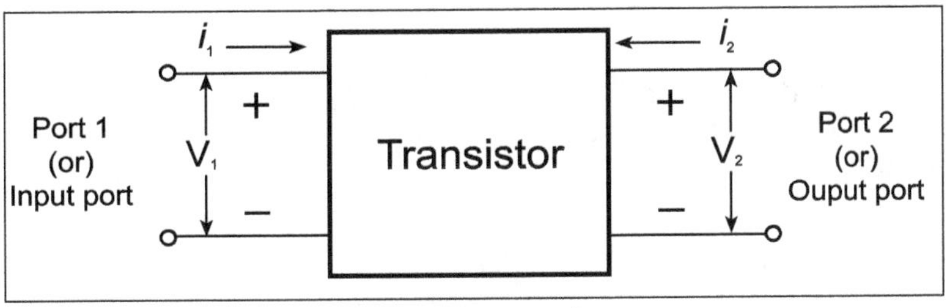

Two port network.

If the input current i_1 and the output voltage v_2 are taken as independent variables, then the input voltage v_1 and output current i_2 is written as:

$$V_1 = A\,V_2 + B(-I_2) \qquad ...(1)$$

$$i_2 = h_{21}i_1 + h_{22}v_2 \qquad ...(2)$$

Four hybrid parameters h_{11}, h_{12}, h_{21} and h_{22} are defined as follows:

$h_{11} = \left[\frac{v_1}{i_1}\right]$ with $v_2 = 0 =$ Input impedance with short circuited output port.

$h_{22} = \left[\frac{i_2}{v_2}\right]$ with $i_1 = 0 =$ Output admittance with open circuited input port.

$h_{12} = \left[\frac{v_1}{v_2}\right]$ with $i_1 = 0 =$ Reverse voltage transfer ratio with open circuited input port.

$h_{21} = \left[\frac{i_2}{i_1}\right]$ with $v_2 = 0 =$ Forward current gain with short circuited output port.

The dimensions of h-parameters are as follows:

$h_{11} - \Omega$

h_{22} – mhos (℧)

h_{12}, h_{21} – Dimensionless

Since the dimensions are unlike (i.e.,) they are hybrid in nature, these parameters are called as hybrid parameters.

An alternative subscript notation recommended by IEEE is commonly used.

i = 11 = Input.

o = 22 = Output.

f = 21 = Forward transfer.

r = 12 = Reverse transfer.

Notations used in Transistor Circuit

In case of transistors, another subscript (b for base, e for emitter or c for collector) is added to indicate the type of configuration.

For example:

$h_{11e} = h_{ie}$ = Short circuit input impedance.

$h_{22e} = h_{oe}$ = Open circuit output admittance.

$h_{12e} = h_{re}$ = Open circuit reverse voltage transfer ratio.

$h_{21e} = h_{fe}$ = Short circuit forward current gain.

Hybrid Model for Two Port Network

Based on the definition of hybrid parameters, the mathematical model for two port networks known as h-parameter is developed as shown in the below figure.

Equations (1) and (2) are written as:

$$v_1 = h_i i1 + h_r v_2$$

$$i_2 = h_f i1 + h_o v_2$$

The model shown above must satisfy these two equations and it can be verified by writing Kirchhoff's voltage law equation for the input loop and KCL equation for the output node. The input circuits have a dependent voltage generator and the output circuit contains a dependent current generator.

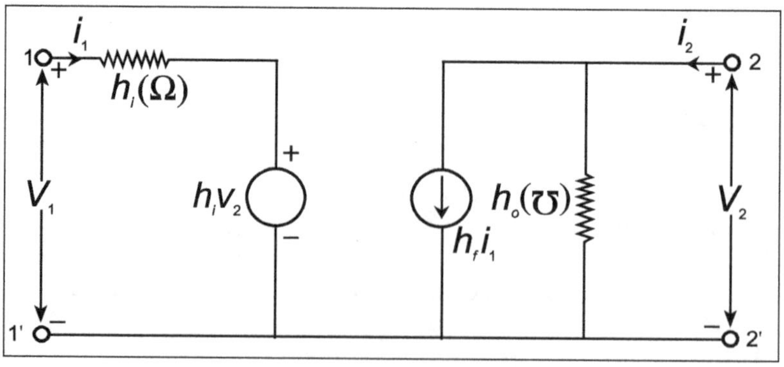

Hybrid model for two port network.

H-parameter Model for CE Configuration

In common emitter (CE) configuration, the emitter of the transistor is common to both the input and output terminals. Base of the transistor is the input terminal and collector of the transistor is the output terminal. The variables are i_B, i_C, v_B (= v_{BE}) and v_C (= V_{CE}). i_B and v_C are considered as independent variables. v_B and i_C are considered as dependent variables.

$$v_B = f_1(i_B, v_C)$$

$$i_C = f_2(i_B, v_C)$$

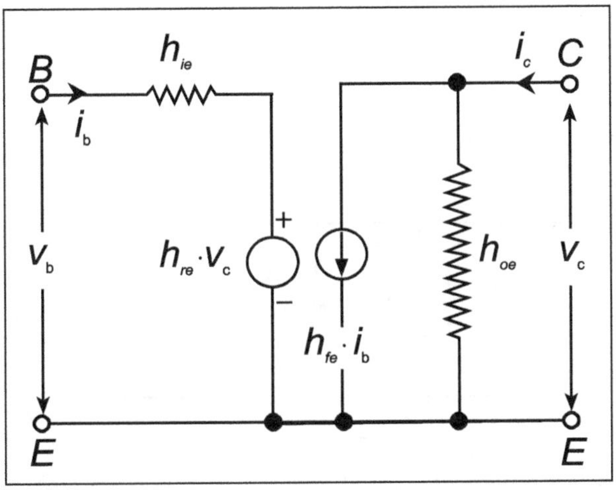

Hybrid model for transistor in CE configuration.

Equations for CE configuration is written as:

$$v_b = h_{ie} i_b + h_{re} v_C$$

$$i_c = h_{fe} i_b + h_{oe} v_C$$

Subscript 'e' indicates that, the h-parameters are used for CE configuration.

h_{ie} = Input impedance when collector emitter terminal is short circuited.

h_{re} = Reverse voltage gain when base emitter terminal is open circuited.

h_{fe} = Forward current gain when collector emitter terminal is short circuited.

h_{oe} = Output admittance when base emitter terminal is open circuited.

Problems

1. Let us find the z-parameters for the network shown in following figure.

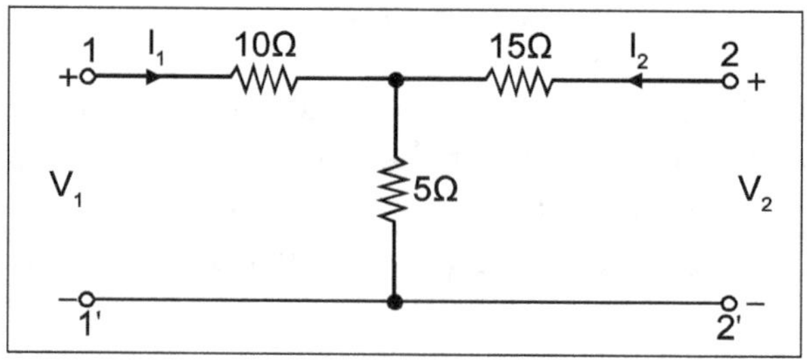

(a)

Solution:

By definition z-parameters are given as:

$$V_1 = z_{11} I_1 + z_{12} I_2$$

$$V_2 = z_{21} I_1 + z_{22} I_2$$

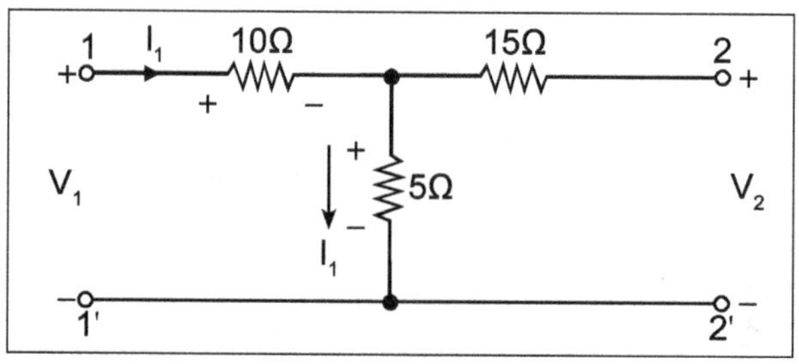

(A) Let $I_2 = 0$, i.e. port-2 is open circuited as shown in the above figure (a).

As port-2 is open circuited, the current flowing through 5 Ω is also I_1. Note that no current will flow through 15 Ω as it is connected to open terminals.

Applying KVL at input side, we get:

$$-10I_1 - 15\, I_1 + V_1 = 0$$

$$V_1 = 15\, I_1$$

$$\therefore\ z_{11} = \left.\frac{V_1}{I_1}\right|_{I_2=0} = 15\Omega$$

From the above figure (a), we can write:

$$V_2 = 5\, I_1$$

$$\therefore\ z_{21} = \left.\frac{V_2}{I_1}\right|_{I_2=0} = 5\Omega$$

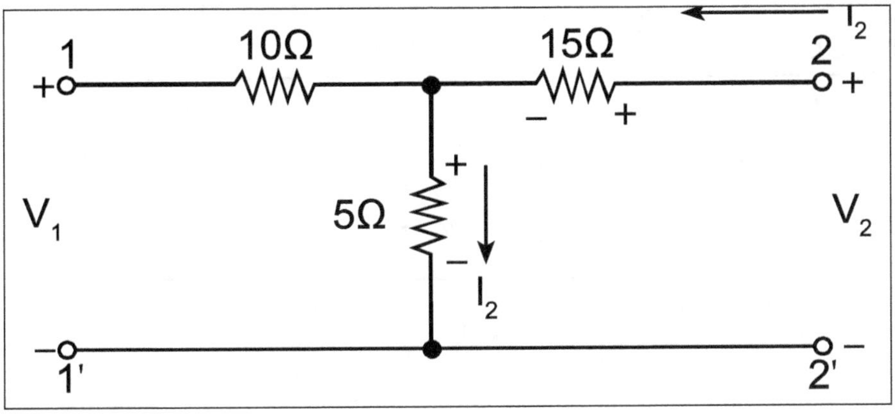

(b)

(B) Let $I_1 = 0$, i.e. port-1 is open circuited as shown in the above figure (b).

As port-1 is open circuited, the current flowing through 5 Ω is also I_2. Note that no current will flow through 10 Ω as it is connected to open terminals.

Applying KVL at output side, we get as:

$$-15I_2 - 5I_2 + V_2 = 0$$

$$\therefore V_2 = 20\, I_2$$

$$\therefore Z_{22} = \left.\frac{V_2}{I_2}\right|_{I_1=0}$$

$$= 20\ \Omega$$

From the above figure, we can write as:

$$V_1 = 5\, I_2$$

$$\therefore Z_{12} = \left.\frac{V_1}{I_2}\right|_{I_1=0} = 5\Omega$$

Hence, z-parameters of the given network are:

$$[z] = \begin{bmatrix} Z_{11} & Z_{12} \\ Z_{21} & Z_{22} \end{bmatrix} = \begin{bmatrix} 15 & 5 \\ 5 & 20 \end{bmatrix} \Omega$$

2. Using the definition, let us find y-parameters of the two port network as shown in the below figure.

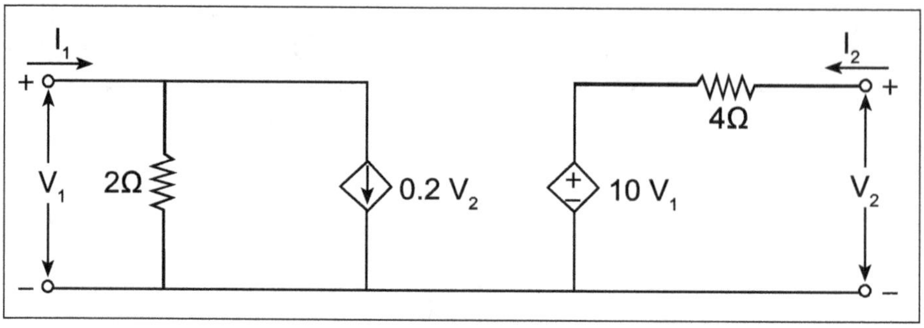

(a)

Solution:

By definition, y-parameters are given as:

$$I_1 = y_{11} V_1 + y_{12} V_2$$

$$I_2 = y_{21} V_1 + y_{22} V_2$$

(i) Let $V_2 = 0$, i.e. port-2 is short circuited as shown in the below figure (a).

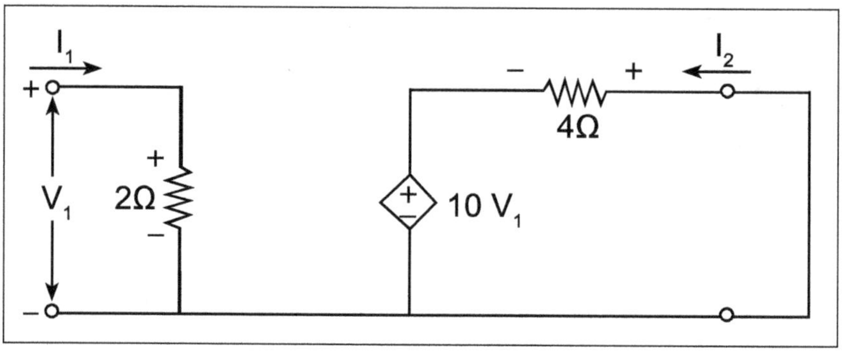

$$\therefore V_1 = 2\,I_1$$

$$\therefore y_{11} = \left.\frac{I_1}{V_1}\right|_{V_2=0} = \frac{1}{2} = 0.5\ \mho$$

Applying KVL, at output side:

$$4I_2 + 10V_1 = 0$$

$$\therefore 4I_2 = -10V_1$$

$$y_{21} = \left.\frac{I_2}{V_1}\right|_{V_2=0} = -2.5\mho$$

(ii) Let V_1 = 0 i.e. port-1 is short circuited. As V_1 = 0, the value of dependent voltage source of value 10 V_1 becomes zero as shown in the below figure (b).

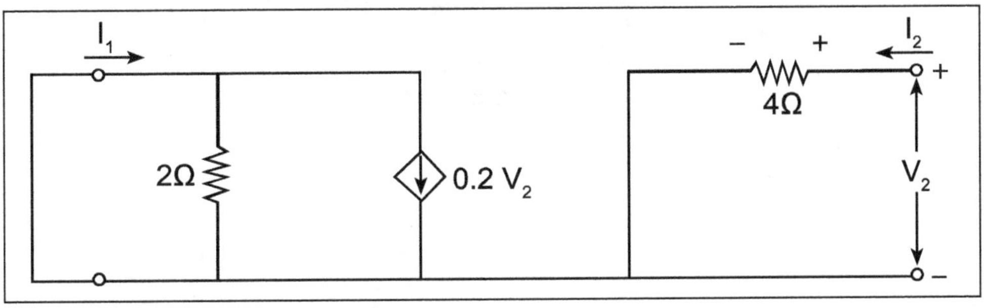

(b)

$$V_2 = 4I_2$$

$$y_{22} = \left.\frac{I_2}{V_2}\right|_{V_1=0} = \frac{1}{4} = 0.25\,\mho$$

Consider the below figure, current source is converted to voltage source.

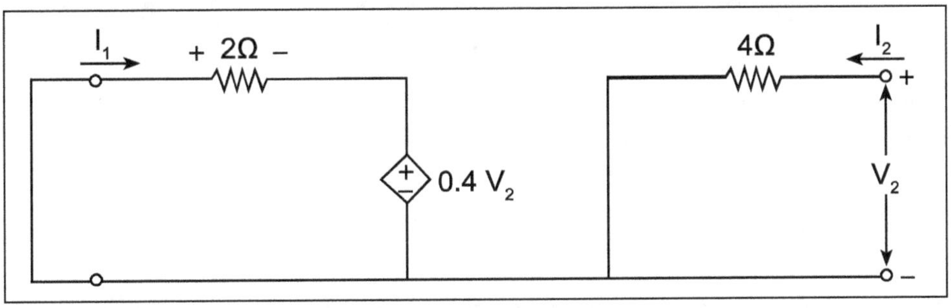

Applying KVL:

$$-2I_1 + 0.4V_2 = 0$$

$$0.4V_2 = 2I_1$$

$$y_{12} = \left.\frac{I_1}{V_2}\right|_{V_1=0} = \frac{0.4}{2} = 0.2\,\mho$$

The y-parameters are given as:

$$[y] = \begin{bmatrix} 0.5 & 0.2 \\ -2.5 & 0.25 \end{bmatrix} \mho$$

3. Let us determine h-parameters after writing transformed network.

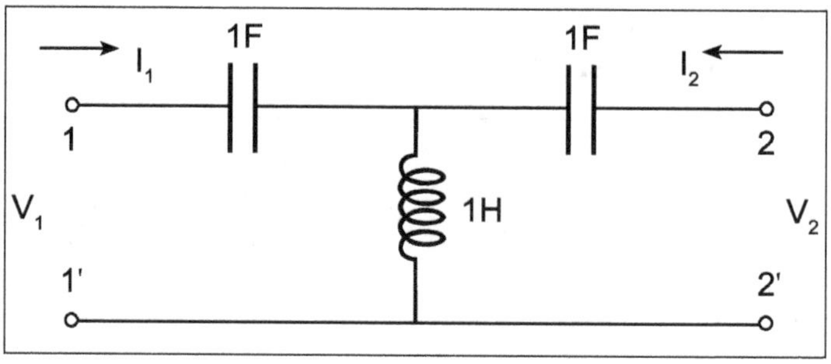

Solution:

The transformed network is as follows, in the below figure (a).

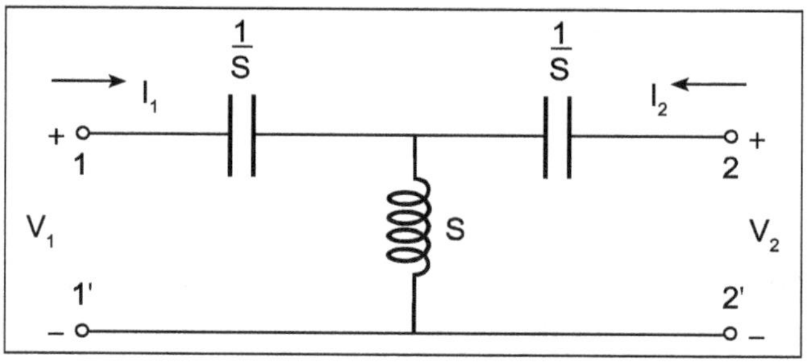

(a)

By definition, h-parameters are given as:

$$V_1 = h_{11} I_1 + h_{12} V_2$$

$$I_2 = h_{21} I_1 + h_{22} V_2$$

(A) Let $V_2 = 0$ i.e. port-2 short circuited as shown in the below figure (b).

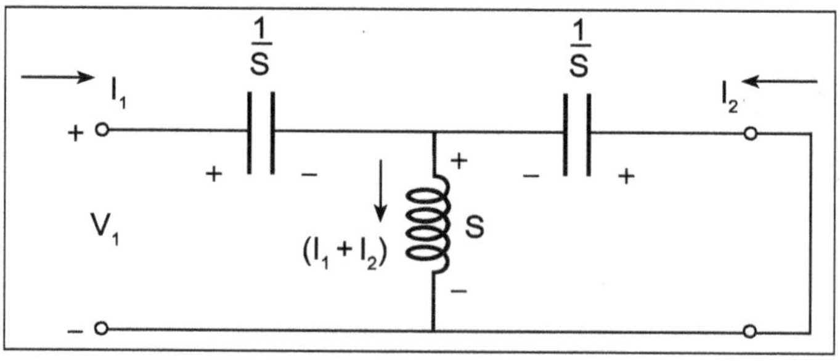

(b)

By current divider rule:

$$I_2 = -I_1 \left[\frac{s}{s + \frac{1}{s}} \right] = -\left[\frac{s^2}{s^2 + 1} \right] I_1 \qquad ...(1)$$

Applying KVL at the input side:

$$-\left(\frac{1}{s}\right)I_1 - s(I_1 + I_2) + V_1 = 0$$

$$\therefore V_1 = \left(\frac{1}{s}\right)I_1 - sI_1 + sI_2$$

$$\therefore V_1 = \left(\frac{1}{s} + s\right)I_1 + sI_2$$

Putting value of I2 from equation (1), we get:

$$V_1 = \left(\frac{1+s^2}{s}\right) + s\left[-\left(\frac{s^2}{s^2+1}\right)\right]I_1$$

$$\therefore \quad V_1 = \frac{(1+s^2)(s^2+1) - s^4}{s(s^2+1)}I_1$$

$$\therefore \quad V_1 = \frac{s^4 + 2s^2 + 1 - s^4}{s(s^2+1)}I_1$$

$$\therefore \quad V_1 = \frac{2s^2+1}{s(s^2+1)}I_1 \qquad \text{...(2)}$$

$$\therefore \quad h_{11} = \left.\frac{V_1}{I_1}\right|_{V_2=0} = \frac{2s^2+1}{s(s^2+1)}\,\Omega$$

From the above equation (1), we can write as:

$$h_{21} = \left.\frac{I_2}{I_1}\right|_{V_2=0} = \frac{s^2}{s^2+1}$$

(B) Let $I_1 = 0$ i.e. port-1 is open circuited as shown in the below figure(c).

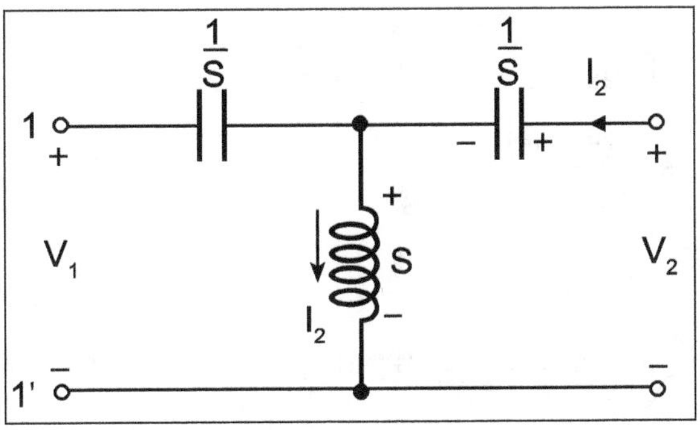

As port 1 is open circuited, the current through inductor is I2. Note that no current will flow through capacitor connected to terminal 1 as it is open circuited.

Applying KVL at the output side, we get:

$$-\left(\frac{1}{s}\right)I_2 - (s)\,I_2 + V_2 = 0$$

$$\therefore\ -\left[\frac{1}{s} + s\right]I_2 = -V_2$$

$$\therefore\ \left(\frac{s^2+1}{s}\right)I_2 = V_2 \qquad \ldots(3)$$

$$h_{22} = \left.\frac{I_2}{V_2}\right|_{I_1=0} = \frac{s}{s^2+1}\ \mho$$

From the above figure, we can write:

$$V_1 = sI_2 \qquad \ldots(4)$$

But from equation (3), $I_2 = \left(\frac{s}{s^2+1}\right)V_2$. Putting value of I_2 in equation (4), we get:

$$V_1 = s\left(\frac{s}{s^2+1}\right)V_2$$

$$h_{12} = \left.\frac{V_1}{V_2}\right|_{I_1=0} = \frac{s^2}{s^2+1}$$

Thus h-parameters of the given network are:

$$[h] = \begin{bmatrix} h_{11} & h_{12} \\ h_{21} & h_{22} \end{bmatrix} = \begin{bmatrix} \frac{2s^2+1}{s\left(s^2+1\right)} & \frac{s^2}{s^2+1} \\ -\frac{s^2}{s^2+1} & \frac{s}{\left(s^2+1\right)} \end{bmatrix}$$

4. Let us find the transmission or general parameters for the circuit shown in the below figure.

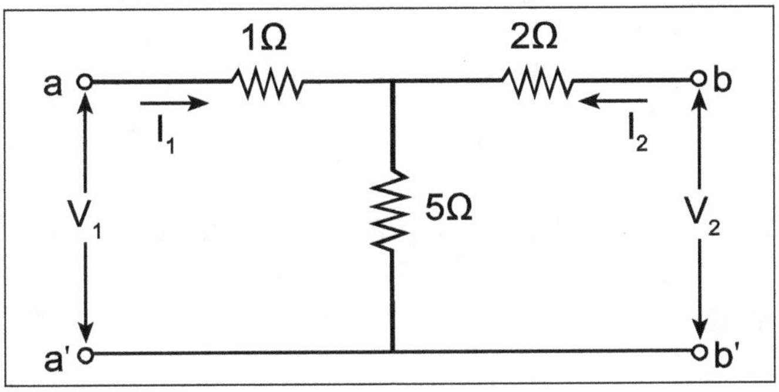

Solution:

By definition, transmission parameters are given by:

$$V_1 = A V_2 + B(-I_2)$$

$$I_1 = C V_2 + D(-I_2)$$

(A) Let $-I_2 = 0$ i.e. open circuit terminals b - b' (i.e. port-2) as shown in the below figure (a).

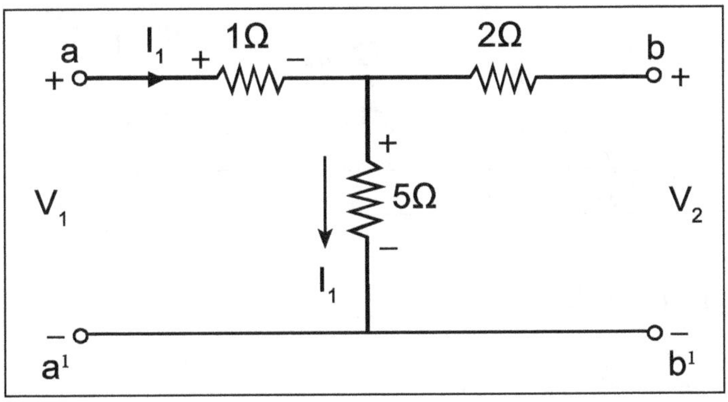

(a)

From the above figure, we can write as:

$$V_2 = 5I_1 \qquad ...(1)$$

$$\therefore\ C = \left.\frac{I_1}{V_2}\right|_{-I_2=0} = \frac{1}{5}\mho$$

Applying KVL at the input side, we get as:

$$-I_1 - 5I_1 + V_1 = 0$$

$$\therefore\ 6I_1 = V_1 \qquad ...(2)$$

From the above equation (1):

$$I_1 = \frac{V_2}{5} \qquad ...(3)$$

Substituting the value of I_1 in the above equation (2), we get as:

$$\therefore\ 6\left(\frac{V_2}{5}\right) = V_1$$

$$\therefore\ V_2 = \frac{5}{6} V_1$$

$$\therefore\ A = \left.\frac{V_1}{V_2}\right|_{-I_2=0} = \frac{6}{5}$$

(B) Let V_2 = 0 i.e. short circuit port-2 (terminals b - b') as shown in the below figure (b).

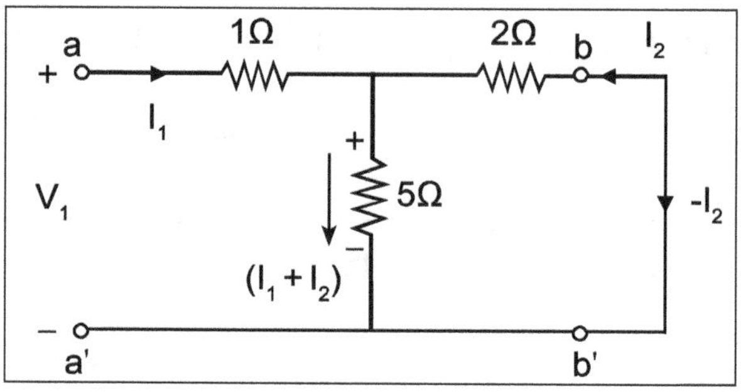

(b)

Applying current divider rule, we get:

$$I_2 = -I_1\left[\frac{5}{2+5}\right]$$

$$\therefore\ -I_2 = I_1\left(\frac{5}{7}\right)$$

$$\therefore\ D = \left.\frac{I_1}{-I_2}\right|_{V_2=0} = \frac{7}{5}$$

Applying KVL at the input side, we get:

$$-I_1 - 5(I_1 + I_2) + V_1 = 0$$

$$\therefore -6I'_1 - 5I_2 = -V_1$$

$$\therefore\ 6I_1 + 5I_2 = V_1$$

$$\therefore 6\left(\frac{-7}{5}I_2\right) + 5I_2 = V_1$$

$$\therefore \left(\frac{-42}{5} + 5\right) I_2 = V_1$$

$$\therefore \frac{-17}{5} I_2 = V_1$$

$$\therefore \left(\frac{17}{5}\right)(-I_2) = V_1$$

$$\therefore B = \left.\frac{V_1}{-I_2}\right|_{V_2=0} = +\frac{17}{5}\Omega$$

Hence, transmission parameters matrix is given by:

$$\begin{bmatrix} A & B \\ C & D \end{bmatrix} = \begin{bmatrix} \frac{6}{5} & \frac{17}{5} \\ \frac{1}{5} & \frac{7}{5} \end{bmatrix}$$

5.2 Interconnection of Two Port Networks, Symmetrical Properties of T and π Networks

When the network is very complex, then it becomes difficult to evaluate two port parameters for such network. Hence such network can be decomposed in small and simple two port network and parameters can be evaluated easily. Depending on the interconnection of these simple two port networks, the parameters of overall network are affected.

Common connections are as follows:

- Series.
- Parallel.
- Cascade.

There are five possible combination of interconnection of two-port networks as follows:

- Cascade connection or tandem connection.
- Parallel connection or parallel-parallel connection.
- Series connection or series-series connection.
- Series-parallel connection.
- Parallel-series connection.

Series Connection or Series-series Connection

When the currents in the series elements are equal, then the voltages add up to give the resultant voltage. It is called as a series connection.

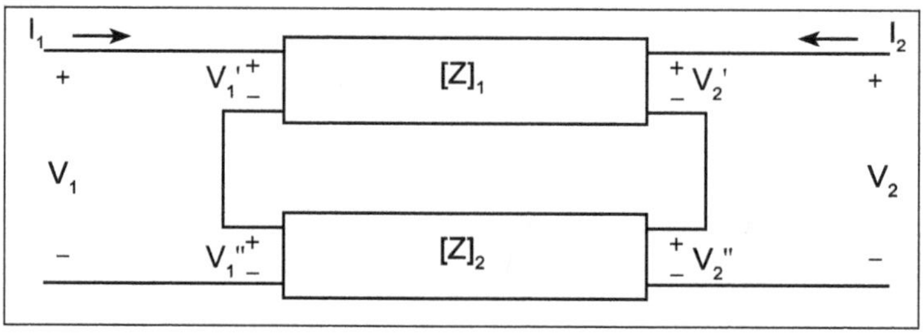

Series-connected two-port network.

Let us consider that the two networks are connected in series as shown in the below figure.

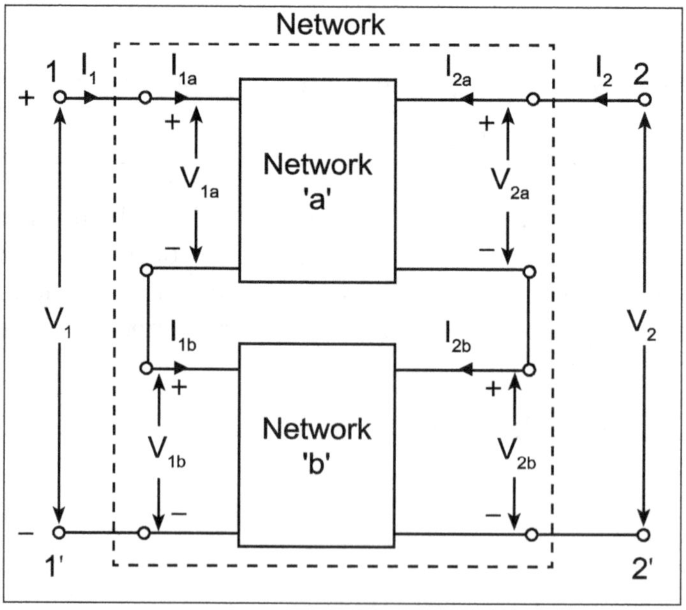

Series connection of two networks.

Z-parameter equations for network 'a' are written as:

$$V_{1a} = Z_{11a} I_{1a} + Z_{12a} I_{2a} \quad ...(1)$$

$$V_{2a} = Z_{21a} I_{1a} + Z_{22a} I_{2a}$$

For network 'b', we can write:

$$V_{1b} = Z_{11b} I_{1b} + Z_{12b} I_{2b} \quad ...(2)$$

$$V_{2b} = Z_{21b} I_{1b} + Z_{22b} I_{2b}$$

From the above figure, voltage V_1 at port 1-1' is equal to the sum of V_{1a} and V_{1b} i.e.

$$V_1 = V_{1a} + V_{1b} \quad ...(3)$$

Substituting V_{1a} and V_{1b} in the above equation (1) and (2), we get:

$$V_1 = Z_{11a} I_{1a} + Z_{11b} I_{1b} + Z_{12a} I_{2a} + Z_{12b} I_{2b} \quad ...(4)$$

$I_1 = I_{1a} = I_{1b}$ since port 1 - 1' is series connected

$I_2 = I_{2a} = I_{2b}$ since port 2 - 2' is series connected

The above equation (4) becomes:

$$V_1 = Z_{11a} I_1 + Z_{11b} I_1 + Z_{12a} I_2 + Z_{12b} I_2$$

$$V_1 = (Z_{11a} + Z_{11b}) I_1 + (Z_{12a} + Z_{12b}) I_2 \quad ...(5)$$

From the above figure, it is clear that the voltage V_2 is equal to sum of V_{2a} and V_{2b} i.e.,

$$V_2 = V_{2a} + V_{2b}$$

$$V_2 = Z_{21a} I_{1a} + Z_{21b} I_{1b} + Z_{22a} I_{2a} + Z_{22b} I_{2b}$$

$$I_1 = I_{1a} = I_{1b} \text{ and } I_2 = I_{2a} = I_{2b}$$

$$V_2 = (Z_{21a} + Z_{21b}) I_1 + (Z_{22a} + Z_{22b}) I_2 \quad ...(6)$$

Z-parameters for the combined network are written as:

$$V_1 = Z_{11} I_1 + Z_{12} I_2$$

$$V_2 = Z_{21} I_1 + Z_{22} I_2$$

Comparing the above two equations with the above equation (5) and (6), we have:

$$Z_{11} = Z_{11a} + Z_{11b} \cdot Z_{12} = Z_{12a} + Z_{12b}$$

$$Z_{21} = Z_{21a} + Z_{21b} \text{ and } Z_{22} = Z_{22a} + Z_{22b} \qquad ...(7)$$

The above equations shows that if the two port networks are series connected, then the Z-parameters of individual two port networks are added.

Parallel Connection or Parallel-parallel Connection

When the voltages in the parallel elements are equal and the currents add up to give the resultant current, then the connection is known as parallel connection.

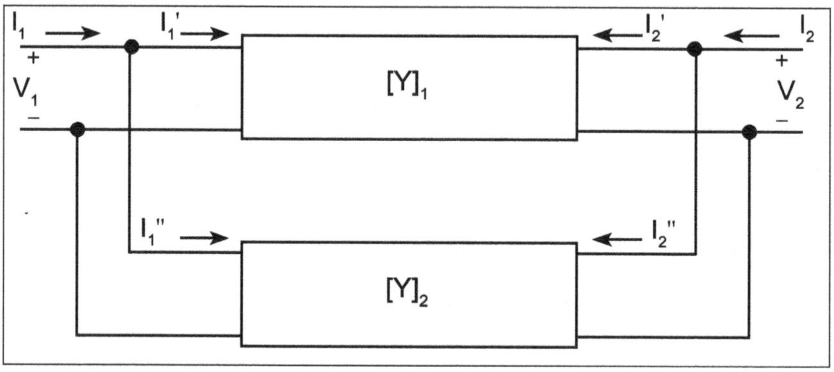

Parallel-connected two-port network.

Let us consider the parallel connected two port networks as shown in the below figure.

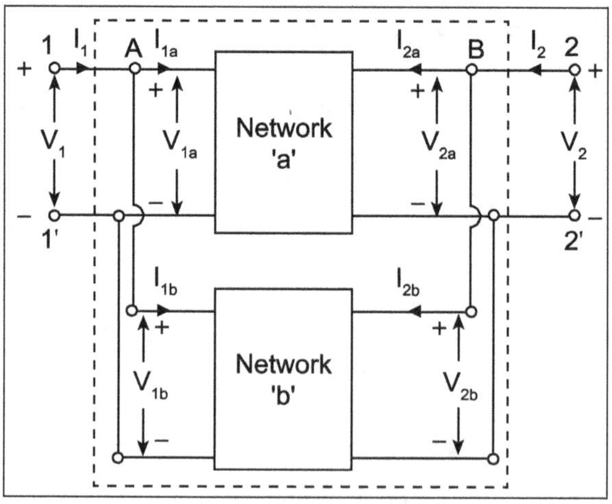

Complex two port network.

For the network 'a', let us write the Y-parameter equations as follows.

$$I_{1a} = Y_{11a} V_{1a} + Y_{12a} V_{2a} \qquad ...(1)$$

$$I_{2a} = Y_{21a} V_{1a} + Y_{22a} V_{2a}$$

Similarly for network 'b', the Y-parameter equations are written as:

$$I_{1b} = Y_{11b} V_{1b} + Y_{12b} V_{2b} \quad ...(2)$$

$$I_{2b} = Y_{21b} V_{1b} + Y_{22b} V_{2b}$$

In the above figure, by applying KCL at node A, we get:

$$I_1 = I_{1a} + I_{1b}$$

$$I_1 = Y_{11a} V_{1a} + Y_{11b} V_{1b} + Y_{12a} V_{2a} + Y_{12b} V_{2b}$$

$$V_1 = V_{1a} = V_{1b} \text{ and } V_2 = V_{2a} = V_{2b}$$

Therefore above equation can be written as:

$$I_1 = (Y_{11a} + Y_{11b}) V_1 + (Y_{12a} + Y_{12b}) V_2 \quad ...(3)$$

If we apply KCL at node B, we get:

$$I_2 = I_{2a} + I_{2b}$$

Substituting I_{2a} and I_{2b}, we get:

$$I_2 = Y_{21a} V_{1a} + Y_{21b} V_{1b} + Y_{22a} V_{2a} + V_{22b} V_{2b}$$

From the above equation (1):

$$V_2 = V_{2a} = V_{2b}$$

$$I_2 = (Y_{21a} + Y_{21b}) V_1 + (Y_{22a} + Y_{22b}) V_2 \quad ...(4)$$

Y-parameter equations for a two port network are given as:

$$I_1 = Y_{11} V_1 + Y_{12} V_2$$

$$I_2 = Y_{21} V_1 + Y_{22} V_2$$

By comparing the above equation (3) and equation (4), we get:

$$Y_{11} = Y_{11a} + Y_{11b}, \ Y_{12a} = Y_{12a} + Y_{12b}$$

$$Y_{21} = Y_{21a} + Y_{21b} \text{ and } Y_{22} = Y_{22a} + Y_{22b} \quad ...(5)$$

The above relationships show that if the two port networks are connected in parallel, then the Y-parameters of the individual networks are added.

Cascade Connection or Tandem Connection

When the output of one network becomes the input to the next network, then the connection is known as cascade connection.

A complex network is built up by cascading simple two port networks. Normal transmission parameter is used in the analysis of cascaded networks.

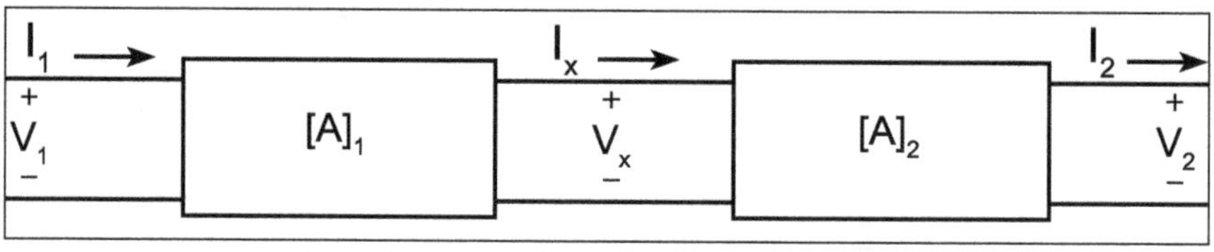

Cascade connection of two-port network.

If two-port networks with z-parameters $[Z]_1, [Z]_2, [Z]_3, ..., [Z]_n$ are connected in series, then the equivalent two-port z-parameters are given as:

$$[Z]_{eq} = [Z]_1 + [Z]_2 + [Z]_3 + ... + [Z]_n \quad ...(1)$$

If two-port networks with y-parameters $[Y]_1, [Y]_2, [Y]_3, ..., [Y]_n$ are connected in parallel, then the equivalent two-port y-parameters are given as,

$$[Y]_{eq} = [Y]_1 + [Y]_2 + [Y]_3 + ... + [Y]_n \quad ...(2)$$

When several two-port networks are connected in cascade and the individual networks have transmission parameters $[A]_1, [A]_2, [A]_3, ..., [A]_n$, then the equivalent two-port parameter will have a transmission parameter as:

$$[A]_{eq} = [A]_1 * [A]_2 * [A]_3 * ... * [A]_n \quad ...(3)$$

Let us consider that two networks are cascaded as shown in the below figure.

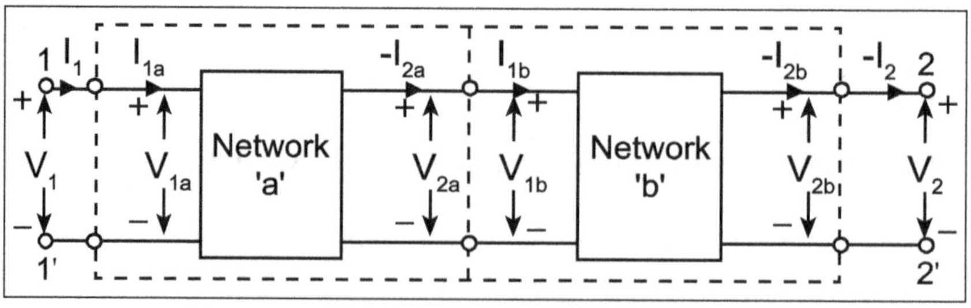

Two networks connected in cascade or tandem.

Let us write the transmission parameters for two networks individually.

For network 'a', we can write transmission parameters in the matrix form as:

$$\begin{bmatrix} V_{1a} \\ I_{1a} \end{bmatrix} = \begin{bmatrix} A_a & B_a \\ C_a & D_a \end{bmatrix} \begin{bmatrix} V_{2a} \\ -I_{2a} \end{bmatrix} \quad ...(1)$$

For network 'b', we can write:

$$\begin{bmatrix} V_{1b} \\ I_{1b} \end{bmatrix} = \begin{bmatrix} A_b & B_b \\ C_b & D_b \end{bmatrix} \begin{bmatrix} V_{2b} \\ -I_{2b} \end{bmatrix} \quad ...(2)$$

From the above figure, we can observe that:

$$I_1 = I_{1a}$$

$$V_1 = V_{1a}$$

$$\begin{bmatrix} V_1 \\ I_1 \end{bmatrix} = \begin{bmatrix} A_a & B_a \\ C_a & D_a \end{bmatrix} \begin{bmatrix} V_{2a} \\ -I_{2a} \end{bmatrix} \quad ...(3)$$

$$V_{2a} = V_{1b}$$

$$-I_{2a} = I_{1b}$$

$$\begin{bmatrix} V_1 \\ I_1 \end{bmatrix} = \begin{bmatrix} A_a & B_a \\ C_a & D_a \end{bmatrix} \begin{bmatrix} V_{1b} \\ -I_{1b} \end{bmatrix}$$

$$\begin{bmatrix} V_1 \\ I_1 \end{bmatrix} = \begin{bmatrix} A_a & B_a \\ C_a & D_a \end{bmatrix} \begin{bmatrix} A_b & B_b \\ C_b & D_b \end{bmatrix} \begin{bmatrix} V_{2b} \\ -I_{2b} \end{bmatrix}$$

From the above figure, we observe that:

$$V_{2b} = V_2$$

$$-I_{2b} = -I_2$$

Hence the above equation becomes:

$$\begin{bmatrix} V_1 \\ I_1 \end{bmatrix} = \begin{bmatrix} A_a & B_a \\ C_a & D_a \end{bmatrix} \begin{bmatrix} A_b & B_b \\ C_b & D_b \end{bmatrix} \begin{bmatrix} V_2 \\ -I_2 \end{bmatrix}$$

By comparing the above equation with standard T-parameter equation, we have:

$$\begin{bmatrix} V_1 \\ I_1 \end{bmatrix} = \begin{bmatrix} A & B \\ C & D \end{bmatrix} \begin{bmatrix} V_2 \\ -I_2 \end{bmatrix}$$

$$\begin{bmatrix} A & B \\ C & D \end{bmatrix} = \begin{bmatrix} A_a & B_a \\ C_a & D_a \end{bmatrix} \begin{bmatrix} A_b & B_b \\ C_b & D_b \end{bmatrix} \quad ...(4)$$

Thus, the overall transmission parameter matrix of cascaded two port networks is obtained by multiplying the transmission parameter matrices of individual two port networks.

Series-parallel Connection

The two two-port networks connected in series-parallel connection is shown in the below figure in which the input ports are connected in series while the output ports are connected in parallel.

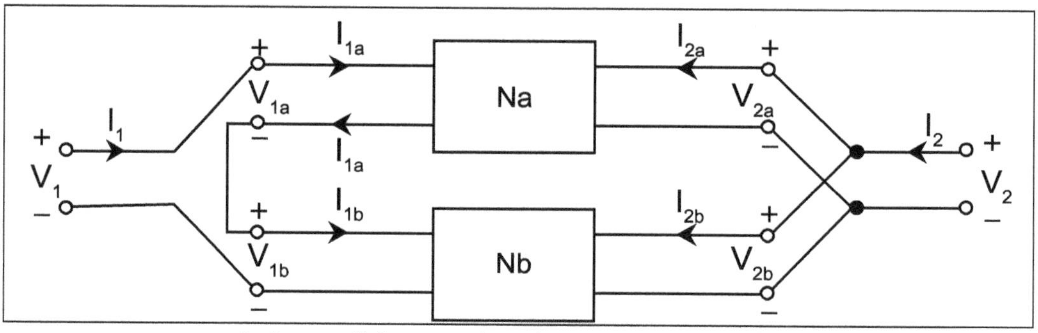

Series parallel connection of two two-port networks.

From the series-parallel connection, we have boundary conditions as follows:

$$V_1 = V_{1a} + V_{1b}$$

$$V_2 = V_{2a} = V_{2b}$$

$$I_1 = I_{1a} + I_{1b}$$

$$I_2 = I_{2a} + I_{2b}$$

For the network N_a, the h-parameter equations are given as:

$$V_{1a} = h_{11a}I_{1a} + h_{12a}V_{2a} \quad ...(1)$$

$$I_{2a} = h_{21a}I_{1a} + h_{22a}V_{2a} \quad ...(2)$$

Similarly, for network N_b, we have:

$$V_{1b} = h_{11b}I_{1b} + h_{12b}V_{2b} \quad ...(3)$$

$$I_{2b} = h_{21b}I_{1b} + h_{22b}V_{2b} \quad ...(4)$$

Adding above equations (1) and (3), we get:

$$V_{1a} + V_{1b} = h_{11a}I_{1a} + h_{12a}V_{2a} + h_{11b}I_{1b} + h_{12b}V_{2b}$$

$$V_1 = (h_{11a} + h_{11b})I_1 + (h_{12a} + h_{12b})V_2 \quad ...(5)$$

$$(\text{Since, } V_1 = V_{1a} + V_{1b}, I_1 = I_{1a} = I_{1b}, V_2 = V_{2a} = V_{2b})$$

Adding the above equations (2) and (4), we get:

$$I_{2a} + I_{2b} = h_{21a}I_{1a} + h_{22a}V_{2a} + h_{21b}I_{1b} + h_{22b}V_{2b}$$

$$I_2 = (h_{21a} + h_{21b})I_1 + (h_{22a} + h_{22b})V_2 \quad ...(6)$$

$$(\text{Since, } I_2 = I_{2a} + I_{2b}, I_1 = I_{1a} = I_{1b}, V_2 = V_{2a} = V_{2b})$$

The h-parameters of the series-parallel connected networks are written in matrix form as:

$$\begin{bmatrix} V_1 \\ I_2 \end{bmatrix} = \begin{bmatrix} h_{11} & h_{12} \\ h_{21} & h_{22} \end{bmatrix} \begin{bmatrix} I_1 \\ V_2 \end{bmatrix} \quad ...(7)$$

The equations (5) and (6) can be written in matrix form as:

$$\begin{bmatrix} V_1 \\ I_2 \end{bmatrix} = \begin{bmatrix} h_{11a} + h_{11b} & h_{12a} + h_{12b} \\ h_{21a} + h_{21b} & h_{22a} + h_{22b} \end{bmatrix} \begin{bmatrix} I_1 \\ V_2 \end{bmatrix}$$

From the above equations (7) and (8), we have,

$$h_{11} = h_{11a} + h_{11b} \qquad h_{12} = h_{12a} + h_{12b}$$

$$h_{21} = h_{21a} + h_{21b} \qquad h_{22} = h_{22a} + h_{22b}$$

$$[h] = [h_a] + [h_b]$$

The overall h-parameters matrix for series-parallel connected two-port networks is the sum of h-parameters matrices of each individual two-port network connected in series-parallel.

This result may be generalized for any number of two-port networks connected in series-parallel.

Parallel-series Connection

The two two-port networks connected in parallel-series connection is shown in the below figure in which the input ports are connected in parallel while the output ports are connected in series.

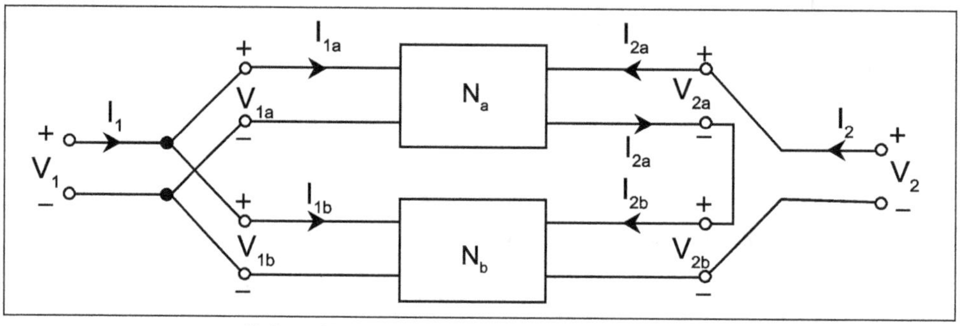

Parallel-series connection of two two-port networks.

From the parallel-series connection, we have boundary conditions as follows:

$$V_1 = V_{1a} = V_{1b} \qquad V_2 = V_{2a} + V_{2b}$$

$$I_1 = I_{1a} + I_{1b} \qquad I_1 = I_{2a} + I_{2b}$$

For the network N_a, the g-parameters equations are given as:

$$I_{1a} = g_{11a} V_{1a} + g_{12a} I_{2a} \qquad \text{...(1)}$$

$$V_{2a} = g_{21a} V_{1a} + g_{22a} I_{2a} \qquad \text{...(2)}$$

Similarly, for network N_b, we have:

$$I_{1b} = g_{11b} V_{1b} + g_{12b} I_{2b} \qquad \text{...(3)}$$

$$V_{2a} = g_{21b} V_{1b} + g_{22b} I_{2b} \qquad \text{...(4)}$$

Adding above equations (1) and (3), we get:

$$I_{1a} + I_{1b} = g_{11a} V_{1a} + g_{12a} I_{2a} + g_{11b} V_{1b} + g_{12b}$$

$$I_1 = (g_{11a} + g_{11b})\, V_1 + (g_{12a} + g_{12b}) I_2 \qquad \text{...(5)}$$

$$\left(\text{Since, } I_1 = I_{1a} + I_{1b}, V_1 = V_{1a} = V_{1b}, I_2 = I_{2a} = I_{2b}\right)$$

Adding above equations (2) and (4), we get:

$$V_{2a} + V_{2b} = g_{21a} V_{1a} + g_{22a} I_{2a} + g_{21b} V_{1b} + g_{22b} I_{2b}$$

$$V_2 = \left(g_{21a} + g_{21b}\right) V_1 + \left(g_{22a} + g_{22b}\right) I_2 \quad ...(6)$$

$$\left(\text{Since, } V_2 = V_{2a} + V_{2b}, V_1 = V_{1a} = V_{1b}, I_2 = I_{2a} = I_{2b}\right)$$

The g-parameters of the parallel -series connected networks is written in matrix form as:

$$\begin{bmatrix} I_1 \\ V_2 \end{bmatrix} = \begin{bmatrix} g_{11} & g_{12} \\ g_{21} & g_{22} \end{bmatrix} \begin{bmatrix} V_1 \\ I_2 \end{bmatrix} \quad ...(7)$$

The equations (5) and (6) are written in matrix form as:

$$\begin{bmatrix} I_1 \\ V_2 \end{bmatrix} = \begin{bmatrix} g_{11a} + g_{11b} & g_{12a} + g_{12b} \\ g_{21a} + g_{21b} & g_{22a} + g_{22b} \end{bmatrix} \begin{bmatrix} V_1 \\ I_2 \end{bmatrix} \quad ...(8)$$

From equations (7) and (8), we have:

$$g_{11} = g_{11a} + g_{11b} \qquad g_{12} = g_{12a} + g_{12b}$$

$$g_{21} = g_{21a} + g_{21b} \qquad g_{22} = g_{22a} + g_{22b}$$

In matrix form, it is given as:

$$[g] = [g_a] + [g_b]$$

The overall g-parameters matrix for parallel-series connected two-port networks is the sum of g-parameters matrices of each individual two-port network connected in parallel-series connection.

This result may be generalized for any number of two-port networks connected in parallel-series.

Problems

1. Let us find the equivalent y-parameters for the bridge T-network shown in the below figure.

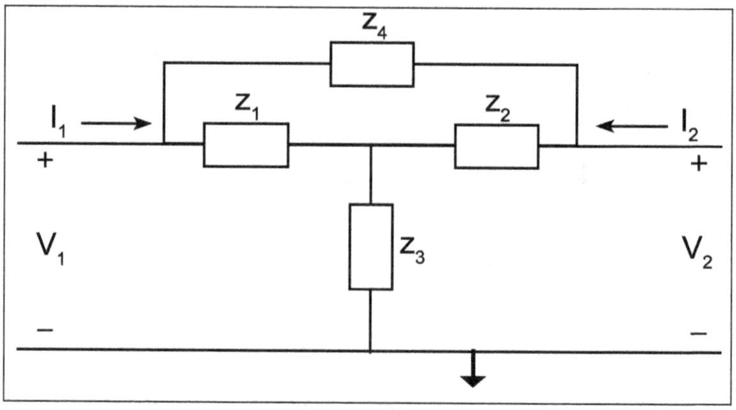

Bridge T network.

Solution:

The bridge T network is redrawn as shown in the below figure.

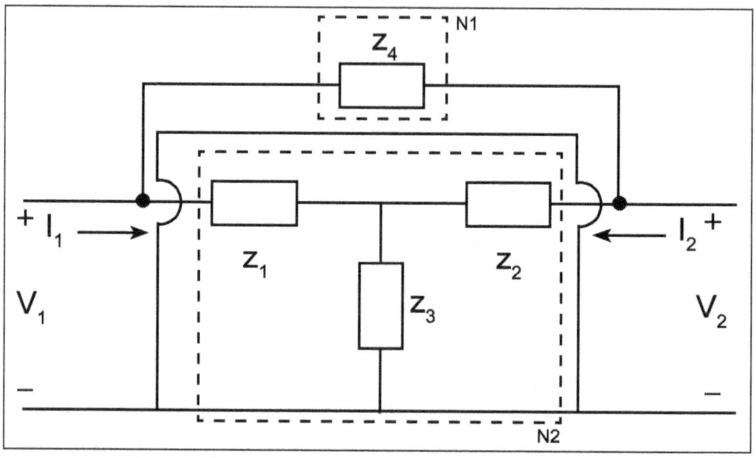

Alternative representation of bridge T network.

The z-parameters of network N_2 are given as:

$$[Z] = \begin{bmatrix} Z_1 + Z_3 & Z_3 \\ Z_3 & Z_2 + Z_3 \end{bmatrix}$$

Let us convert the z-parameters to y-parameters as follows:

$$y_{11} = \frac{Z_2 + Z_3}{Z_1 Z_2 + Z_1 Z_3 + Z_2 Z_3}$$

$$y_{12} = \frac{-Z_3}{Z_1 Z_2 + Z_1 Z_3 + Z_2 Z_3}$$

$$y_{21} = \frac{-Z_3}{Z_1 Z_2 + Z_1 Z_3 + Z_2 Z_3}$$

$$y_{22} = \frac{Z_1 + Z_3}{Z_1 Z_2 + Z_1 Z_3 + Z_2 Z_3} \quad ...(1)$$

Transmission parameters of network N_1 is given as follows:

$$a_{11} = 1 \quad a_{12} = Z_4$$
$$a_{21} = 0 \quad a_{22} = 1$$

Let us convert the transmission parameters to y-parameters as follows:

$$y_{11} = \frac{1}{Z_4}$$

$$y_{12} = -\frac{1}{Z_4}$$

$$y_{21} = -\frac{1}{Z_4}$$

$$y_{22} = \frac{1}{Z_4} \quad ...(2)$$

The equivalent y-parameters of the bridge T-network is given as follows.

$$y_{11eq} = \frac{1}{Z_4} + \frac{Z_2 + Z_3}{Z_1 Z_2 + Z_1 Z_3 + Z_2 Z_3}$$

$$y_{12eq} = -\frac{1}{Z_4} - \frac{Z_3}{Z_1 Z_2 + Z_1 Z_3 + Z_2 Z_3}$$

$$y_{21eq} = -\frac{1}{Z_4} - \frac{Z_3}{Z_1 Z_2 + Z_1 Z_3 + Z_2 Z_3}$$

$$y_{22eq} = \frac{1}{Z_4} + \frac{Z_1 + Z_3}{Z_1 Z_2 + Z_1 Z_3 + Z_2 Z_3} \quad ...(3)$$

2. Let us find the transmission parameters of the below figure.

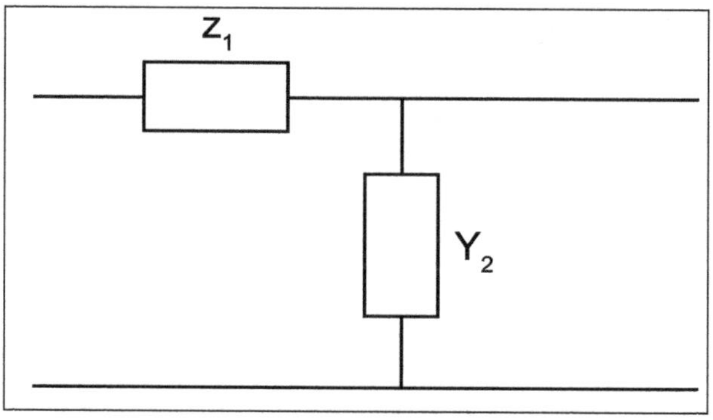

Cascade network.

Solution:

The above figure can redraw as shown in the below figure.

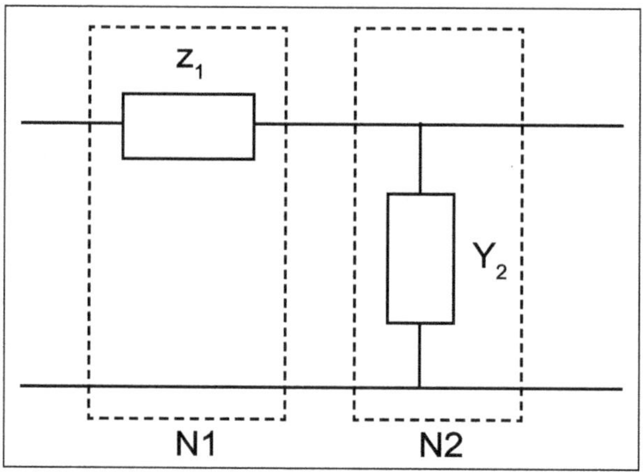

Cascade of two networks N_1 and N_2.

The transmission parameters of network N_1 are given as:

$$a_{11} = 1 \quad a_{12} = Z_1$$
$$a_{12} = 0 \quad a_{22} = 1$$

The transmission parameters of network N_2 are given as:

$$a_{11} = 1 \quad a_{12} = 0$$
$$a_{21} = Y_2 \quad a_{22} = 1$$

The transmission parameters of the above figure are given as:

$$\begin{bmatrix} a_{11} & a_{12} \\ a_{21} & a_{22} \end{bmatrix}_{eq} = \begin{bmatrix} 1 & Z_1 \\ 0 & 1 \end{bmatrix} \begin{bmatrix} 1 & 0 \\ Y_2 & 1 \end{bmatrix} = \begin{bmatrix} 1+Z_1 Y_2 & Z_1 \\ Y_2 & 1 \end{bmatrix}$$

3. Let us find the transmission parameters for the cascaded system shown in the below figure where the resistance values are in ohms.

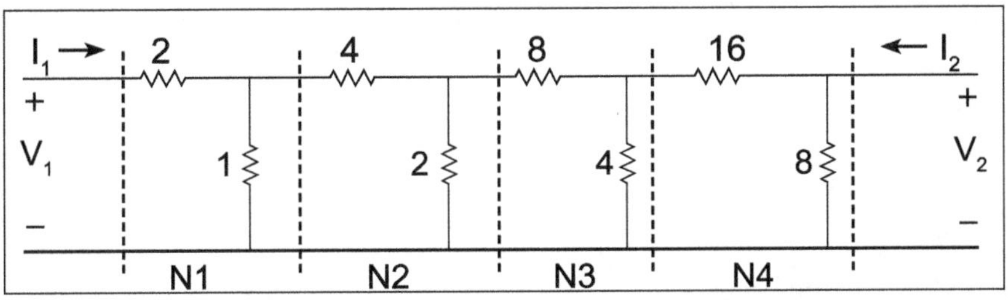

Solution:

The above figure is considered as four networks N_1, N_2, N_3 and N_4 which are connected in cascade.

The transmission parameters of the above figure are given as:

$$[a]_{N1} = \begin{bmatrix} 3 & 2 \\ 1 & 1 \end{bmatrix}$$

$$[a]_{N2} = \begin{bmatrix} 3 & 4 \\ 0.5 & 1 \end{bmatrix}$$

$$[a]_{N3} = \begin{bmatrix} 3 & 8 \\ 0.25 & 1 \end{bmatrix}$$

$$[a]_{N4} = \begin{bmatrix} 3 & 16 \\ 0.125 & 1 \end{bmatrix}$$

The transmission parameters of the above figure is obtained by using the following MATLAB program.

MATLAB Script

```
diary ex7_9.dat

%Transmission parameters of caseded network

a1 =[3 2; 1 1];

a2 =[3 4; 0.5 1];

a3 =[3 8; 0.25 1];
```

```
a4 =[3 16; 0.125 1];

% equivalent transmission parameter

a=a1*(a2*(a3*a4))

diary
```

The value of matrix a is

$$a = \begin{bmatrix} 112.2500 & 630.0000 \\ 39.3750 & 221.0000 \end{bmatrix}$$

4. Let us calculate T-parameters for the network shown in the below figure using cascade connection of networks.

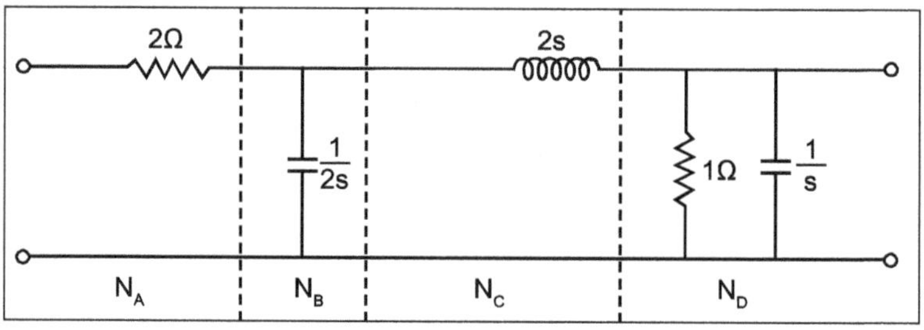

Solution:

Step 1:

T-parameters equation:

$$V_1 = AV_2 - BI_2 \quad \text{...(1)}$$

$$I_1 = CV_2 - DI_2 \quad \text{...(2)}$$

Step 2:

T-parameters for N_A section:

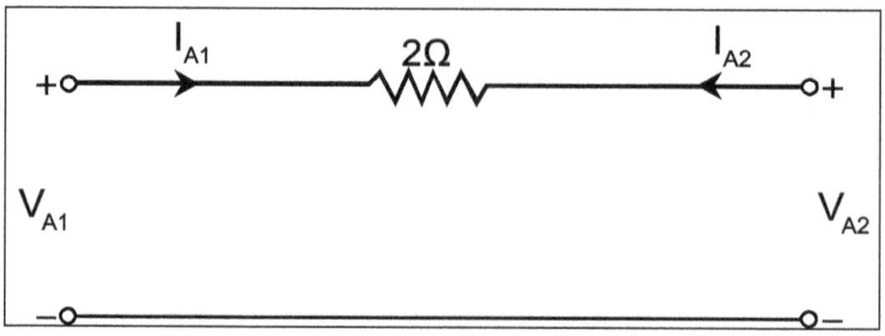

For the above network, we can write:

$$2I_{A1} + V_{A2} - V_{A1} = 0$$

$$I_{A1} = -I_{A2}$$

$$V_{A1} = V_{A2} - 2I_{A2} \qquad ...(3)$$

$$I_{A1} = 0V_{A2} - I_{A2} \qquad ...(4)$$

By comparing the above equations (1), (3) and (2), (4), then:

$$[T_A] = \begin{bmatrix} 1 & 2 \\ 0 & 1 \end{bmatrix}$$

Step 3:

T-parameters for N_B section:

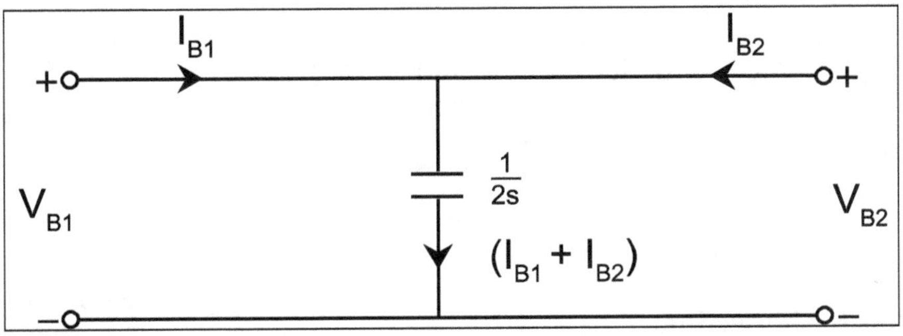

By KVL, we can write as:

$$V_{B1} = \frac{1}{2s}(I_{B1} + I_{B2}) = V_{B2}$$

$$V_{B1} = V_{B2} - 0I_{B2} \qquad ...(5)$$

$$I_{B1} = 2s\, V_{B2} - I_{B2} \qquad ...(6)$$

By comparing the above equations (1), (5) and (2), (6), we get:

$$[T_B] = \begin{bmatrix} 1 & 0 \\ 2s & 1 \end{bmatrix}$$

Step 4:

T-parameters for N_C section:

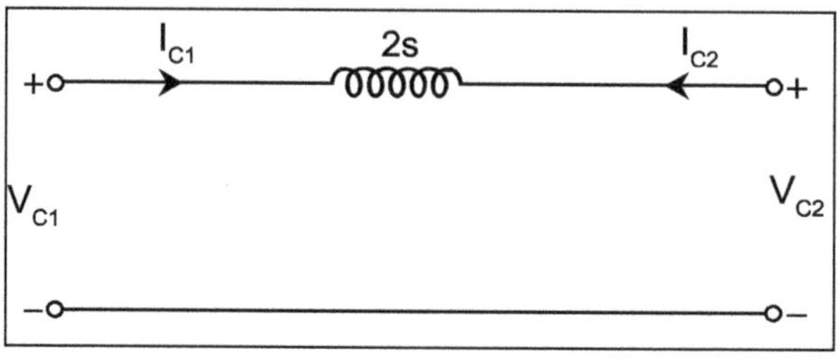

From the above circuit, we can write:

$$2sIC_1 + V_{C2} - V_{C1} = 0 \quad \text{and} \quad I_{C1} = -I_{C2}$$

$$V_{C1} = V_{C2} + 2sI_{C1}$$

$$V_{C1} = V_{C2} - 2sI_{C2} \qquad ...(7)$$

$$I_{C1} = 0.V_{C2} - I_{C2} \qquad ...(8)$$

By comparing the above equations (1), (7) and (2), (8), we get:

$$[T_C] = \begin{bmatrix} 1 & 2s \\ 0 & 1 \end{bmatrix}$$

Step 5:

T-parameters for N_D section:

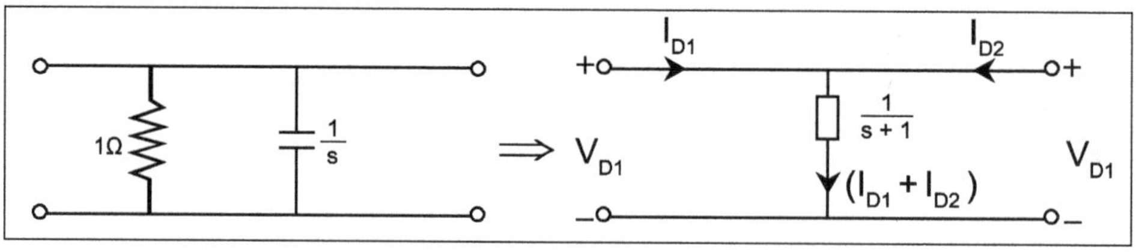

From the above circuit, we can write:

$$V_{D1} = V_{D2} = \left(\frac{1}{s+1}\right)(I_{D1} + I_{D2})$$

$$V_{D1} = V_{D2} - 0.I_{D2} \qquad ...(9)$$

$$I_{D1} = (s+1)V_{D2} - I_{D2} \qquad ...(10)$$

By comparing the above equations (1), (9) and (2), (10), we get:

$$[T_D]=\begin{bmatrix} 1 & 0 \\ s+1 & 1 \end{bmatrix}$$

Step 6:

Overall T-parameters for given cascade connected network:

For cascade connection, the overall T-parameters in matrix form is given as:

$$[T]=[T_A][T_B][T_C][T_D]$$

$$\therefore [T]=\begin{bmatrix} 1 & 2 \\ 0 & 1 \end{bmatrix}\begin{bmatrix} 1 & 0 \\ 2s & 1 \end{bmatrix}\begin{bmatrix} 1 & 2s \\ 0 & 1 \end{bmatrix}\begin{bmatrix} 1 & 0 \\ s+1 & 1 \end{bmatrix}$$

$$=\begin{bmatrix} 1+4s & 2 \\ 2s & 1 \end{bmatrix}\begin{bmatrix} 1+2s^2+2s & 2s \\ s+1 & 1 \end{bmatrix}=\begin{bmatrix} 8s^3+10s^2+8s+3 & 8s^2+2s+2 \\ 4s^3+4s^2+3s+1 & 4s^2+1 \end{bmatrix}$$

5. Let us determine the g-parameters for the network shown in the below figure.

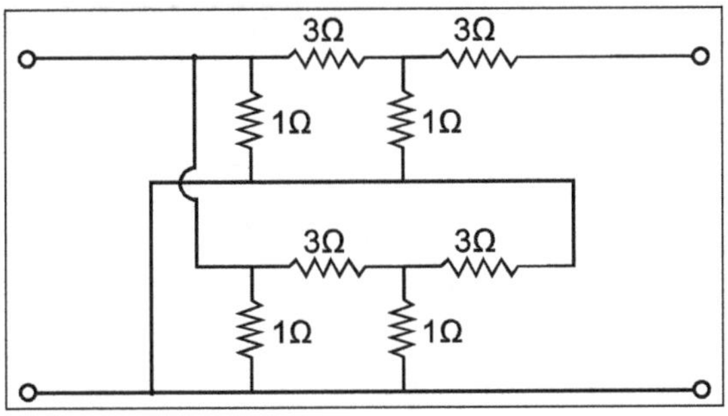

Solution:

Step 1:

Interconnections of network:

The given network is parallel-series connection of two identical networks. The identical network is shown in the below figure.

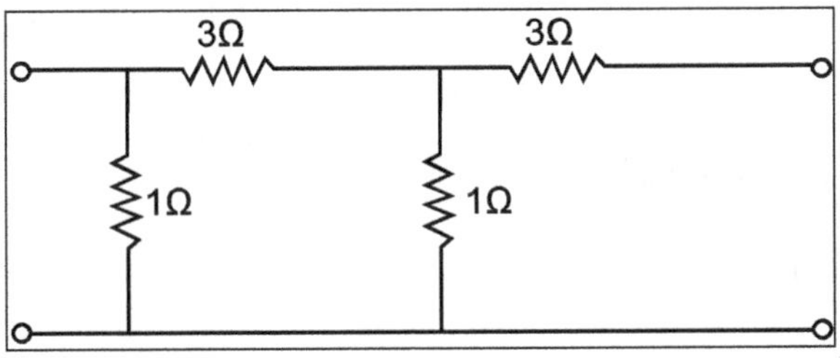

Step 2:

g-parameters equation:

$$I_1 = g_{11} V_1 + g_{12} I_2 \quad ...(1)$$

$$V_2 = g_{21} V_1 + g_{22} I_2 \quad ...(2)$$

Step 3:

Distribution of current in each branch:

Let us write the KVL equations in identical network.

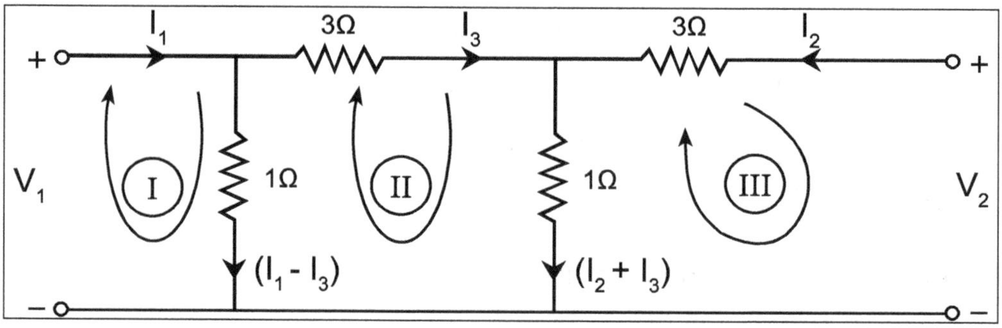

By KVL, we can write as:

Loop 1:

$$1.(I_1 - I_2) - V_1 = 0$$

$$V_1 = I_1 - I_3 \quad ...(3)$$

Loop 2:

$$3.I_3 + 1.(I_2 + I_3) - 1.(I_1 - I_3) = 0$$

$$5I_3 = I_1 - I_2 \quad ...(4)$$

Loop 3:

$$V_2 - 1.(I_2 + I_3) - 3I_2 = 0$$

$$V_2 = 4I_2 + I_3 \quad ...(5)$$

Substituting I_3 from equation (4) in (3), we get:

$$V_1 = I_1 - \frac{1}{5}(I_1 - I_2)$$

$$V_1 = \frac{4}{5}I_1 + \frac{1}{5}I_2 \qquad ...(6)$$

Putting I_3 from equation (4) in (5), we get:

$$V_2 = 4I_2 + \frac{1}{5}(I_1 - I_2)$$

$$V_2 = \frac{1}{5}I_1 + \frac{19}{5}I_2 \qquad ...(7)$$

$$I_1 = \frac{5}{4}V_1 - \frac{1}{4}I_2 \qquad ...(8)$$

$$V_1 - 4V_2 = \frac{1}{5}I_2 - \frac{76}{5}I_2$$

$$-4V_2 = -V_1 - 15I_2$$

$$V_2 = \frac{1}{4}V_1 + \frac{15}{4}I_2 \qquad ...(9)$$

Step 4:

g-parameters for identical networks:

By comparing the above equations (1) and (8), we get:

$$g_{11} = \frac{5}{4}, g_{12} = -\frac{1}{4}$$

By comparing the above equations (2) and (9), we get:

$$g_{21} = \frac{1}{4}, g_{22} = \frac{15}{4}$$

In matrix form, it is given as:

$$[g_a] = \begin{bmatrix} \frac{5}{4} & \frac{-1}{4} \\ \frac{1}{4} & \frac{15}{4} \end{bmatrix}$$

Step 5:

Overall g-parameters for a given network:

For parallel-series connection, we have:

$$[g]=[g_a]+[g_b]$$

For identical networks:

$$[g_a]=[g_b]$$

$$[g]=2[g_a]=\begin{bmatrix} \frac{5}{2} & -\frac{1}{2} \\ \frac{1}{2} & \frac{15}{2} \end{bmatrix}$$

Permissions

All chapters in this book are published with permission under the Creative Commons Attribution Share Alike License or equivalent. Every chapter published in this book has been scrutinized by our experts. Their significance has been extensively debated. The topics covered herein carry significant information for a comprehensive understanding. They may even be implemented as practical applications or may be referred to as a beginning point for further studies.

We would like to thank the editorial team for lending their expertise to make the book truly unique. They have played a crucial role in the development of this book. Without their invaluable contributions this book wouldn't have been possible. They have made vital efforts to compile up to date information on the varied aspects of this subject to make this book a valuable addition to the collection of many professionals and students.

This book was conceptualized with the vision of imparting up-to-date and integrated information in this field. To ensure the same, a matchless editorial board was set up. Every individual on the board went through rigorous rounds of assessment to prove their worth. After which they invested a large part of their time researching and compiling the most relevant data for our readers.

The editorial board has been involved in producing this book since its inception. They have spent rigorous hours researching and exploring the diverse topics which have resulted in the successful publishing of this book. They have passed on their knowledge of decades through this book. To expedite this challenging task, the publisher supported the team at every step. A small team of assistant editors was also appointed to further simplify the editing procedure and attain best results for the readers.

Apart from the editorial board, the designing team has also invested a significant amount of their time in understanding the subject and creating the most relevant covers. They scrutinized every image to scout for the most suitable representation of the subject and create an appropriate cover for the book.

The publishing team has been an ardent support to the editorial, designing and production team. Their endless efforts to recruit the best for this project, has resulted in the accomplishment of this book. They are a veteran in the field of academics and their pool of knowledge is as vast as their experience in printing. Their expertise and guidance has proved useful at every step. Their uncompromising quality standards have made this book an exceptional effort. Their encouragement from time to time has been an inspiration for everyone.

The publisher and the editorial board hope that this book will prove to be a valuable piece of knowledge for students, practitioners and scholars across the globe.

Index

www.ingramcontent.com/pod-product-compliance
Lightning Source LLC
LaVergne TN
LVHW080636210926
848485LV00003B/271